T0295343

An Introduction to Experimental Nuclear Reactions

An Introduction to Experimental Nuclear Reactions

Chinmay Basu

CRC Press is an imprint of the
Taylor & Francis Group, an **informa** business

First Edition published 2022
by CRC Press
6000 Broken Sound Parkway NW, Suite 300, Boca Raton, FL 33487-2742

and by CRC Press
2 Park Square, Milton Park, Abingdon, Oxon, OX14 4RN

CRC Press is an imprint of Taylor & Francis Group, LLC

Library of Congress Cataloging-in-Publication Data
Names: Basu, Chinmay, 1967- author.
Title: An introduction to experimental nuclear reactions / Chinmay Basu.
Description: First edition. | Boca Raton ; CRC Press, 2022. | Summary: "Introduction to Experimental Nuclear Reactions is a book with a concise and simple approach on the subject of experimental nuclear physics. The subject being very technical is dealt with in a lucid way so that the reader can grasp the concept and later gain hands on experience while doing field work. In this book, theoretical, experimental and instrumentation aspects are all covered with emphasis on accelerator based techniques which forms the basis of the subject of experimental nuclear physics"-- Provided by publisher.
Identifiers: LCCN 2021027900 (print) | LCCN 2021027901 (ebook) | ISBN 9780367539634 (hardback) | ISBN 9780367539665 (paperback) | ISBN 9781003083863 (ebook)
Subjects: LCSH: Nuclear physics--Experiments. | Nuclear physics--Fieldwork. | Nuclear reactions.
Classification: LCC QC786.75 .B37 2022 (print) | LCC QC786.75 (ebook) | DDC 539.7/50724--dc23
LC record available at https://lccn.loc.gov/2021027900
LC ebook record available at https://lccn.loc.gov/2021027901

ISBN: 978-0-367-53963-4 (hbk)
ISBN: 978-0-367-53966-5 (pbk)
ISBN: 978-1-003-08386-3 (ebk)

DOI: 10.1201/9781003083863

Typeset in Minion Pro
by SPi Technologies India Pvt Ltd (Straive)

Contents

List of Figures, ix
Preface, xi
About the Author, xv

Chapter 1 ▪ Basic Concepts 1

1.1 INTRODUCTION TO NUCLEAR REACTIONS 1
1.2 THE NUCLEUS 3
1.3 BASIC DEFINITION OF A NUCLEAR REACTION 7
1.4 EXPERIMENTAL SCENARIO 9

Chapter 2 ▪ Cross-Section 11

2.1 EXPERIMENTAL AND THEORETICAL CROSS-SECTIONS 11
2.2 PARTIAL AND TOTAL CROSS-SECTIONS 13
2.3 INCLUSIVE AND EXCLUSIVE CROSS-SECTIONS 14
2.4 SYSTEM OF REFERENCES-LABORATORY AND CENTRE OF MASS 15

Chapter 3 ▪ Kinematics 19

3.1 TWO-BODY KINEMATICS 19
3.2 THREE-BODY KINEMATICS 22
3.3 APPLICATION OF KINEMATICS FOR SAMPLE REACTION 23

Chapter 4 ▪ Types of Nuclear Reactions 25
4.1 CLASSIFICATION OF REACTIONS 25
4.2 SCATTERING 28
4.3 DIRECT REACTION 29
4.4 COMPOUND AND PRECOMPOUND REACTIONS 31
4.5 ASTROPHYSICAL REACTIONS 34

Chapter 5 ▪ Nuclear Reaction Experiments 41
5.1 BASIC FACILITIES REQUIRED FOR EXPERIMENTAL NUCLEAR REACTION STUDIES 41
5.2 LARGE SCATTERING CHAMBER EXPERIMENT 43
5.3 SMALL CHAMBER EXPERIMENT 48
5.4 EXPERIMENTS WITH SEPARATORS AND SPECTROMETERS 48
5.5 EXPERIMENTAL SETUPS FOR ASTROPHYSICAL REACTIONS 49

Chapter 6 ▪ Accelerators for Nuclear Reactions 51
6.1 BRIEF HISTORY OF ACCELERATORS 51
6.2 BASIC DEFINITIONS AND CLASSIFICATION OF ACCELERATORS 52
6.3 CHARGED PARTICLE ACCELERATORS 53
6.4 NEUTRON GENERATORS 59
6.5 ION OPTICS AND BEAM DIAGNOSTIC DEVICES 60

Chapter 7 ▪ Vacuum Techniques 63
7.1 BASIC DEFINITIONS AND CLASSIFICATION 63
7.2 VACUUM PUMPS 65
7.3 VACUUM MEASURING/CONTROLLING DEVICES 70

Chapter 8 ▪ Radiation Detectors 75
8.1 CLASSIFICATION AND TYPES OF RADIATION DETECTORS 75
8.2 DETECTOR EFFICIENCY AND RESOLUTION 77

8.3 GAS DETECTORS 78
8.4 SOLID STATE DETECTORS 84

CHAPTER 9 ■ Electronics and Data Acquisition System 91

9.1 ANALOG PULSE FORMATION FROM RADIATION DETECTORS 91
9.2 ANALOG TO DIGITAL CONVERSION 95
9.3 BASIC ELECTRONIC CIRCUITS 96
9.4 SPECIALISED ELECTRONIC CIRCUITS 99

CHAPTER 10 ■ Introduction to the Concept of Errors 103

10.1 STATISTICAL, SYSTEMATIC AND PROGRESSIVE ERRORS 103
10.1.1 Statistical Errors 103
10.1.2 Systematic Errors 104
10.1.3 Progressive Errors 105
10.2 CONFIDENCE LIMIT 106
10.3 REDUCTION OF ERRORS IN NUCLEAR REACTION MEASUREMENTS WITH EXAMPLES 106

CHAPTER 11 ■ Theoretical Models 109

11.1 INTRODUCTION TO DIFFERENT REACTION MODELS 109
11.2 BASIC THEORY OF TRANSMISSION AND SCATTERING THROUGH A BARRIER 110
11.2.1 Direct Reaction Models 113
11.3 COMPOUND NUCLEAR REACTION MODEL 122
11.4 THEORIES OF ASTROPHYSICAL REACTIONS 125
11.5 PRE-COMPOUND REACTION MODELS 126
11.6 INTRODUCTION TO SOME AVAILABLE PROGRAMS 127

INDEX, 129

List of Figures

1.1	Plum Pudding model of the atom.	2
1.2	Bohr's planetary model of the atom.	3
1.3	Binding energy per nucleon plot.	4
1.4	Neutron separation energy as a function of nucleon number.	5
1.5	Rotational (a) and vibrational (b) spectra of nuclei.	6
1.6	Single-particle spectrum of a nucleus.	6
1.7	Line of stability and driplines.	7
3.1	Variation of α particle energy with respect to emission angle in $^{19}F(p, \alpha)$ reaction.	23
3.2	Variation of energy of any two of the three particles emitted in $d+^{19}F$ reaction.	24
4.1	Elastic and inelastic scattering cross-section.	29
4.2	Discrete states populated in a nuclear reaction.	30
4.3	Energy spectra for (a) compound nucleus and precompound (b) direct (c) breakup reaction; angular distribution in (c) direct (d) and (e) compound emission; excitation function for (e) non-resonant and (f) resonant reaction.	34
4.4	Gamow peak (solid line), fusion excitation function (dashed line), and Maxwell–Boltzmann energy distribution (dotted line).	37
5.1	A multi-detector array.	43
5.2	A large scattering chamber.	44
5.3	TOF setup.	45
5.4	ΔE-E particle identifier spectrum.	47
6.1	Tandem accelerator.	52
6.2	Van de Graaff and Pelletron accelerators.	54
6.3	Cockroft–Walton generator (voltage doubler).	55
6.4	Cyclotron.	57
6.5	Quadrupole.	60
6.6	Einzel lens.	61
6.7	Dipole magnet.	61
7.1	Oil rotary pump in one stage.	66
7.2	Oil rotary pump in next stage of compression.	67
7.3	Roots pump.	68

7.4 Scroll pump. 69
7.5 Diffusion pump. 69
7.6 Turbomolecular pump. 71
7.7 A basic vacuum generation setup. 72
8.1 Basic design of a gas detector. 80
8.2 Different operating regions of a gas detector. 80
8.3 Transverse and axial field geometry. 82
8.4 Single-wire gas detector. 82
8.5 Multi-wire gas detector. 83
8.6 Metal (Au)-semiconductor (Si) before contact. 87
8.7 Metal (Au)-semiconductor (Si) Schottky contact. 87
8.8 Metal (Al)-semiconductor (Si) before contact. 88
8.9 Metal (Al)-semiconductor (Si) Ohmic contact. 88
9.1 Basic forms of different electronic setup. 97
9.2 Electronic setup for two-detector coincidence. 98
9.3 Leading edge discriminator. 99
9.4 Position sensing by charge division method. 100
10.1 Reduction of statistical error. 107
10.2 Uncertainty or error in energy measurement. 108

Preface

There are a number of books on nuclear reactions published in recent years. Most of these books present an explicit treatment on the theory of nuclear reactions. Some of these books focus on a particular type of reaction mechanism. However, none of these books treat the field from an experimental perspective. Some books have been written on radiation detectors and their associated instrumentation. However, nuclear reaction studies essentially require an understanding of ion beam accelerator studies.

This book introduces a beginner to the basic setup of an accelerator-based experiment. Chapter 1 is a primary introduction to the concepts of nuclear reaction. The cross-section is the main observable which is defined in various ways. This is important, as it helps one to understand different reaction mechanisms in a convenient way. For example, angular distribution, which is a graph of a partial cross-section with an angle of the emitted particle, is useful for the identification of the reaction mechanism. On the other hand, the excitation function, which is a plot of the total cross-section with the incident energy, is useful for studying resonances. Cross-sections are discussed in detail in Chapter 2. Kinematics of two-body and three-body final-state reactions represents an essential knowledge for understanding nuclear reaction experiments. The knowledge of kinematics helps in setting up the detectors in the experiment. For example, in the elastic scattering of two similar nuclei, it is useless to put a detector beyond a certain scattering angle. This is possible to know if one understands the kinematics of elastic scattering. Kinematics is described in Chapter 3. This chapter also describes the two systems of reference, namely the laboratory system (pertinent for an experiment) and centre of mass frame (pertinent for calculations), and their transformations.

The different types of reactions are then introduced in Chapter 4, with reference to their experimental signatures. The reactions are classified into two-body and three-body types as they are commonly studied. Some exotic reactions, such as those involving radioactive nuclei with a comparison to reactions with loosely bound stable nuclei, are discussed. The low-energy astrophysical reactions are briefly discussed, as well as the indirect methods that are useful to study astrophysical reactions at low energies.

The basic setup of a nuclear reaction experiment is described in Chapter 5. The basic setup requires an ion beam accelerator facility to energize the projectile, a scattering chamber facility, target preparation, detector setup, electronics and data acquisition system. Important related instruments include vacuum pumps, charge and current measuring devices, and ion-optical elements, among others. Besides the scattering chamber setups, magnetic spectrographs and large detector arrays are also introduced. Though this chapter discusses the experimental facilities, the descriptions mainly focus on the underlying physical principals rather than on the technical aspects.

Chapter 7 deals with the basic vacuum techniques required for doing an experiment. The vacuum chambers are discussed with reference to a large-diameter scattering chamber for charged particle measurements and small-sized target chambers for detecting gamma emissions. Principles of some commonly used pumps and vacuum measuring and leak detection devices are described. Radiation detectors are discussed in great detail in many books, but here a compact description is presented based on utility. Energy and position measurements are described separately with simple detector configurations. The solid-state detectors involve both the semiconductor and scintillation-type detectors. In comparison to solid-state detectors, gaseous detectors and their basic construction and principle of operation are discussed. Basic aspects, such as resolution, efficiency, interaction mechanism of various nuclear particles and radiations utilised in various detector materials and medium, are discussed in Chapter 8.

Chapter 9 deals with the basic electronics required for processing the signal from radiation detectors. The different NIM electronics setups for energy, timing and position measurement are discussed. The basic principle of each electronic module is explained, as well as the nature of signals and their properties. The different types of cables and connectors are also mentioned. The high-density VME and fastbus and modular multiparameter data acquisition systems are explained. In any experimental measurement, errors need to be estimated. In a nuclear reaction experiment, the sources of errors are mainly systematic and statistical in nature. The sources of these errors and how they can be reduced are described.

The procedure to express the experimental errors and the procedure to estimate errors of theoretical calculation are discussed in Chapter 10. The data is generally represented in a graphical form and as described in the chapter on cross-sections. The analysis of the results requires the basic knowledge of the theoretical models, namely the scattering theory, Green's function method, coupled channels method for direct reactions, the Hauser–Feshbach and Weiskopf–Ewing theory for non-resonant compound nuclear reaction, and R-matrix theory for the resonant compound nuclear reaction. Some basic models for describing fusion are also described. Hybrid and exciton models for pre-equilibrium emissions are briefly mentioned. Some of the computer programs available for calculations are also described.

In summary, the book does not focus on a particular subject or detailed technical aspects of nuclear reaction experiments, but gives an overview of the subject with a focus on the underlying physics. The target audience is beginners in the field of nuclear reactions.

About the Author

Chinmay Basu is a professor in the Nuclear Physics Division at the Saha Institute of Nuclear Physics, Kolkata, India. He received his Master's in Physics from the University of Calcutta with a specialization in nuclear physics, and PhD in Physics from the Saha Institute of Nuclear Physics on non-equilibrium nuclear reactions. After his PhD, he continued as a research associate at the same institute and later joined as a faculty member in 2002. His interests in research are nuclear reactions, nuclear astrophysics and development of nuclear gas detectors. He has about 40 publications in reputed international journals and has delivered lectures at various institutes in India and abroad. He is also a guest teacher at the University of Calcutta and regularly teaches in the post-MSc course at the Saha Institute of Nuclear Physics. The author's present interests are in the field of nuclear astrophysics and its study of very-low-energy high-current ion beams. He is currently in charge of FRENA, the accelerator facility at the Saha Institute of Nuclear Physics.

CHAPTER 1

Basic Concepts

1.1 INTRODUCTION TO NUCLEAR REACTIONS

The atomic theory put forward by John Dalton described all matter as made up of indivisible constituents called atoms. The atoms were considered electrically neutral, as otherwise touching materials would lead to electrical shock. However, the observation of cathode rays from materials under the influence of an electric field lead to the discovery that all atoms contain a negatively charged particle – an electron. The electron is too light to constitute the mass of the atom. So J J Thompson proposed the Plum Pudding model (Figure 1.1) of the atom where the electrons acted as plums in a pudding which has almost all the mass of the atom. The pudding must have an equal and opposite positive charge to that of the electrons so that they cancel to make the atom electrically neutral. An experiment by Moseley showed that different elements are characterised by the number of electronic charge in it. So the positive matter should have an equal charge (but of opposite polarity) and much larger mass.

Moseley's experiment showed that hydrogen has only one electron and only one such positive particle with mass almost that of the atom. In order to decipher the distribution of this heavier positively charged particle, Ernest Rutherford devised an experiment with helium nuclei emitted from radioactive decays. Rutherford's experiment was to bombard the gold atoms. He thought that if Thompson's Plum Pudding model were true,

DOI: 10.1201/9781003083863-1

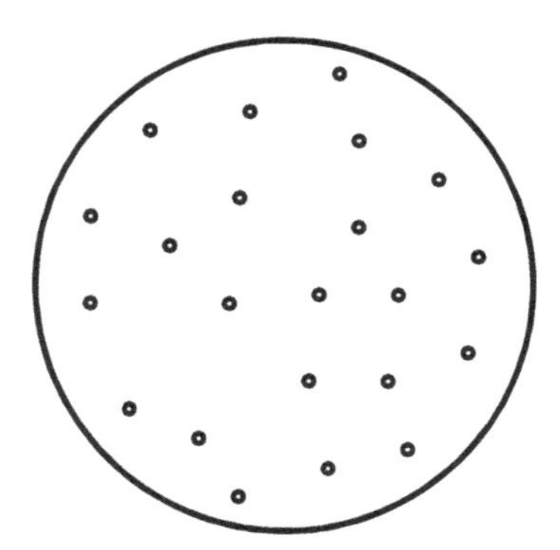

FIGURE 1.1 Plum Pudding model of the atom.

then the alpha would reflect back with a large probability, as the whole of the gold volume would have an even distribution of the positive pudding. This reflection would be as a result of the Coulomb repulsion between the positive puddings of the helium and gold atoms.

But the result of the experiment showed that the probability was very small. This led to the conclusion that the positive matter is very small in a spatial dimension, and Rutherford named this the atomic nucleus. The nuclear charge radius was measured by Hofstadter from electron scattering and was found to be on the order of few Fermi (1 Fermi = 10^{-15} m).

However, the mass of the atom could not be explained by the protons alone. Later, using bombardment by alpha particles of Boron atoms a massless particle was discovered that was as massive as the proton. This is the neutron, and together with the proton they constitute most of the atomic mass. The electrons could not collapse onto the nucleus, as the deBroglie wavelength of an electron confined to the nuclear dimension requires the level of energy that is not available from the electron–proton potential energy. Thus the electrons are far away from the nucleus. Niels Bohr laid down his planetary model of the atom (Figure 1.2).

The planetary model resembles our solar system with the nucleus replicating the sun and electrons the planets. The properties of nucleus showed that the neutrons and protons in it are bound by a short-range attractive nuclear force. It took almost 100 years of research to decipher many properties of the nucleus. Some important properties of the nucleus are discussed in the next section.

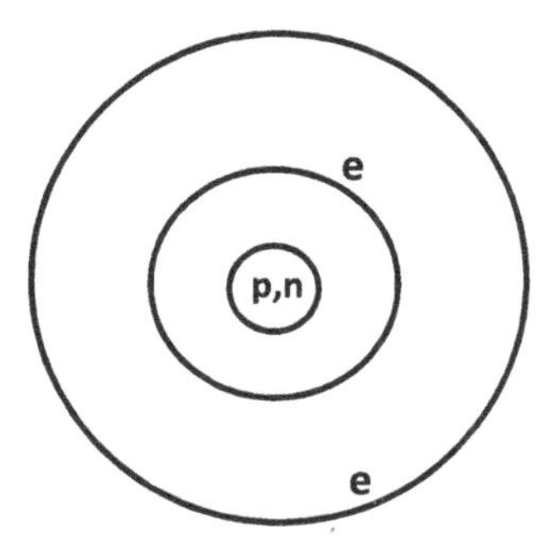

FIGURE 1.2 Bohr's planetary model of the atom.

1.2 THE NUCLEUS

A nucleus ${}^{A}_{Z}X$ of an atomic element is composed of Z protons and N neutrons bound by the strong nuclear force. The properties of the element are determined by the number of electrons and hence by the number of protons that are equal in number. The number of protons is known as the atomic number, and the total number of neutrons and protons constitute the mass number A. The mass $M(A,Z)$ of the nucleus X is not the sum of masses of the neutrons and protons, as some energy is spent in binding the nucleons (the collective name of neutrons and protons). Thus,

$$E_{total} = M(A,Z)c^2 = ZM_pc^2 + NM_nc^2 - B(A,Z) \tag{1.1}$$

where $B(A,Z)$ is the binding energy required to unite the nucleons. In practice, the atomic mass $AM(A,Z)c^2$ has been measured using the relation between atomic and nuclear mass, i.e.

$$AM(A,Z)c^2 = M(A,Z)c^2 + Zm_ec^2 - B_e \tag{1.2}$$

where B_e is energy required to bind the electrons to the nucleus (but at a separation) and m_ec^2 is the mass of an electron. It is to be noted that m_ec^2 is 0.5 MeV and B_e is of the order of eV whereas the masses are of the order of 1,000 MeV. Thus nuclear mass can be approximated from the atomic mass. From Equation 1.1, the binding energy of a nucleus can be calculated. If we plot the ratio of $B(A,Z)$ to A against, we get a very informative graph (Figure 1.3).

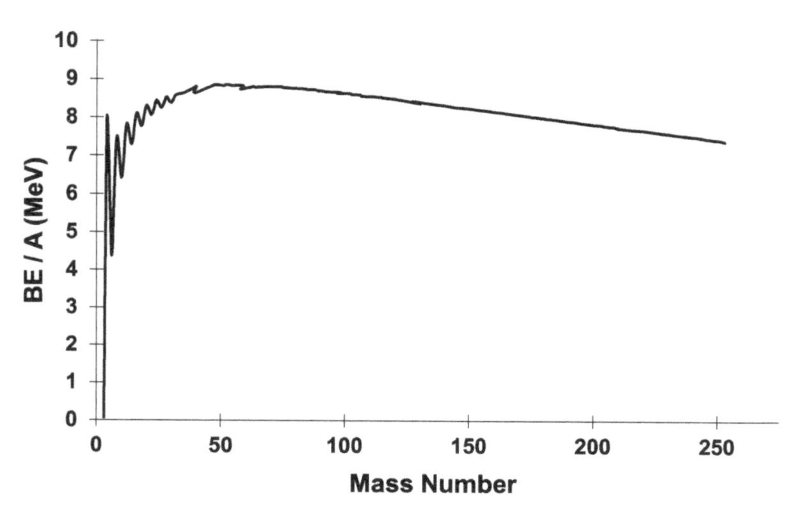

FIGURE 1.3 Binding energy per nucleon plot.

The ratio increases with increasing value of A (values of A in the smaller range) and saturates after a certain value of A. After $A = 56$, the ratio gradually decreases with the increasing value of A. The low A behaviour indicates that if two nuclides are combined to form a nuclide of higher A, the process is more favourable. This is because in the low A region the binding energy increases with increasing A. This means that nuclei with higher A have lower energy, as more energy is lost in binding its nucleons. But this feature is reversed after $A = 56$ when increasing A reduces the binding energy. So the nucleus in this region tends to break up into nuclei of lower mass, as decreasing mass decreases its total energy. Some nuclei have exceptionally high binding energy compared to their neighbouring nuclei and are known as magic nuclei. Nuclei with neutrons or protons equal to 2, 8, 20, 50, 82 show such behaviour.

Mayer suggested the shell picture for the nucleons to explain the magic behaviour of certain nuclei. The shells or energy levels for the nucleons are obtained by solving the Schrodinger equation using a harmonic oscillator potential. The larger shell gaps after N or Z = 2, 8, 20, 28, 50, 82 makes these nuclei require more energy to emit particles and change to a different nuclei. As such they have larger proton (S_p) or neutron separation energies (S_n) compared to their neighbouring nuclei (Figure 1.4).

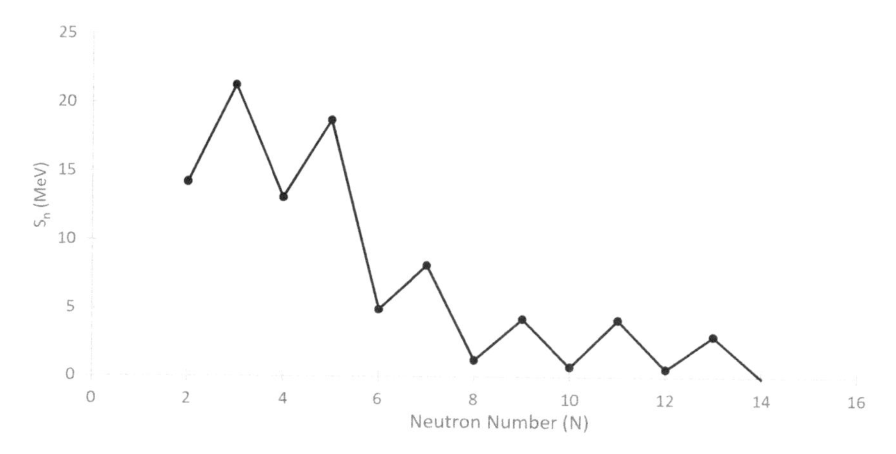

FIGURE 1.4 Neutron separation energy as a function of nucleon number.

$$S_N = M_n c^2 + M\left(A, Z\right)c^2 - M\left(A+1, Z\right)c^2 \tag{1.3}$$

$$S_Z = M_p c^2 + M\left(A, Z\right)c^2 - M\left(A+1, Z+1\right)c^2 \tag{1.4}$$

The shell model can predict the ground state spin and parity of a nucleus with one or few nucleons outside closed shells. However, for shell-closed magic nuclei, the spin is always zero, which is a manifestation of pairing interaction in the nuclear force. If the nucleons do not have sufficient energy to escape the nucleus, it can reside in a higher energy shell, leading to single-particle nuclear excitation. A shell model is a single-particle model, as it uses the one-body Schrodinger equation. A shell-closed or magic nucleus is spherical and cannot rotate. This is because a spherical body has an infinite number of rotation axes, and one cannot define rotation for such a system. However, it can vibrate about its mean spherical shape, resulting in dynamic deformation. Rotational and vibrational modes are shown in Figure 1.5(a) and (b).

A larger number of nucleons outside the closed shell makes the nucleus permanently deformed. Such nuclei can rotate about an axis. Both vibrational and rotational excitations in nuclei are manifested by a very regular energy distribution of the nuclear levels. This is in contrast to the irregular level spacings of single-particle levels (Figure 1.6).

Some nuclei show an alpha cluster structure besides the nucleon structure. The alpha cluster structure is generally observed for energy states close to but below the alpha separation energy of that nucleus. These alpha

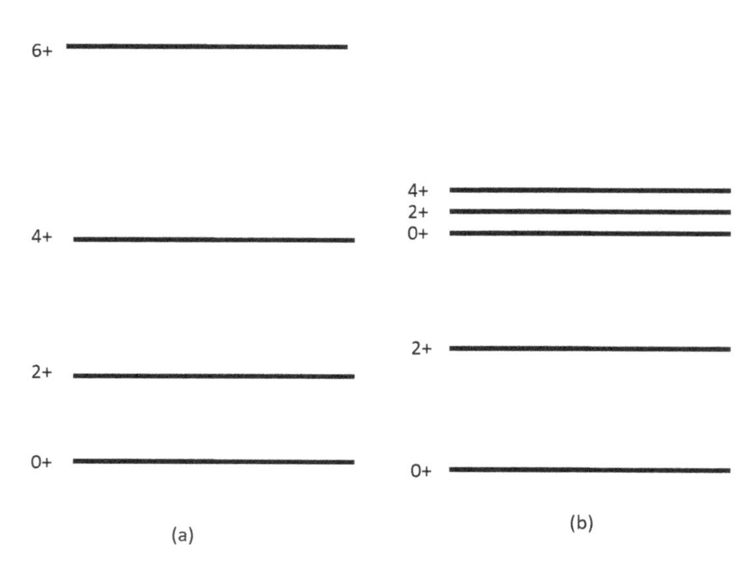

FIGURE 1.5 Rotational (a) and vibrational (b) spectra of nuclei.

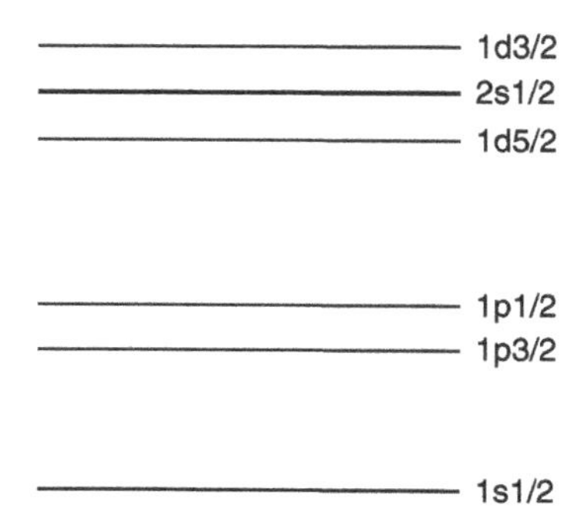

FIGURE 1.6 Single-particle spectrum of a nucleus.

clustering states play important role in helium burning reactions in stars, as they have a greater probability of alpha particle fusion. Recently alpha clustering is observed in a heavier nucleus ^{212}Po. Heavy and superheavy nuclei ($Z>100$) tend to break up in large fragments (spontaneous fission) or alpha particles to attain lower energy configurations. The structure of these different nuclei can be probed through nuclear reactions.

Radioactive (neutron-rich or proton-rich) nuclei are those isotopes that are not stable and decay by emitting alpha, beta and gamma emissions without any external energy. In some nuclei (such as ^{252}Cf), spontaneous fission occurs. A schematic plot (Figure 1.7) of the N v/s Z of different

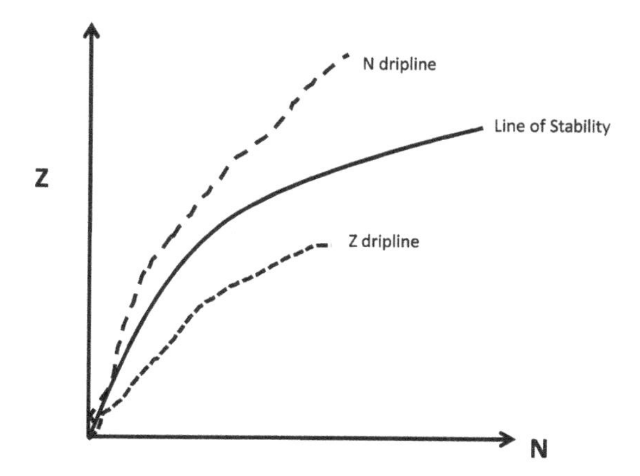

FIGURE 1.7 Line of stability and driplines.

nuclei gives a nice idea about the stability of nuclei as a function of neutron and proton number. The lighter nuclei are stable at, and so the line of stability makes an angle of 45 degrees with the vertical and horizontal axis. At heavier mass, the stability line satisfies $N = Z$. This is to overcome the increasing Coulomb repulsion by the increased nuclear interaction due to the larger number of neutrons. Examples of light stable nuclei are ^{4}He, ^{16}O, and ^{40}Ca that have equal number of protons and neutrons, whereas heavier stable nuclei are ^{120}S, and ^{208}Pb that have a large difference in neutron and proton numbers. The dotted lines in the figure are the neutron drip line, and the dashed line id the proton drip line. Beyond these lines addition of a proton or neutron is not possible, as they cannot be bound with the nucleus to which they are added.

1.3 BASIC DEFINITION OF A NUCLEAR REACTION

A nuclear reaction is an interaction between two nuclei $^{A_1}_{Z_1}P$ and $^{A1}_{z1}T$ such that a different set of nuclei are produced. The simplest of these is a two-nuclei or two-body final state, i.e.

$$^{A1}_{Z1}P + ^{A2}_{Z2}T \rightarrow ^{A3}_{Z3}E + ^{A4}_{Z4}R \tag{1.5}$$

In some situations there may be one or three nuclei in the final state. In case of one nucleus an electromagnetic radiation is associated with it. However, the formation of more than three nuclei is possible at higher energies. In any case, the total energy, atomic number, mass number, spin, parity of the initial and final state must be the same. Conserving the total energy, we can write for the above equation:

$$M_P c^2 + M_T c^2 = M_E c^2 + M_R c^2 \quad (1.6)$$

where $M_i c^2$ and $i = P, T, E, R$ is the relativistic mass. The total energy can be written as the sum of rest mass energy and kinetic energy, i.e.

$$E = M_i c^2 = M_{0i} c^2 + \left(M_i c^2 - M_{0i} c^2\right) \quad (1.7)$$

In the non-relativistic limit, the kinetic energy $(M_i - M_{0i})c^2$ can be approximated as $\frac{1}{2} m v_i^2$. So

$$M_P c^2 + T_P + M_T c^2 + T_T = M_E c^2 + T_T + M_R c^2 + T_R \quad (1.8)$$

Defining a quantity called Q value as

$$Q = \left(M_P + M_T - M_E - M_R\right)c^2 = T_E + T_R - T_P - T_T \quad (1.9)$$

The value of Q may be positive or negative. If the initial pair of nuclei are at rest and masses are such that Q is negative, then the reaction is not energetically possible, as

$$|Q| = T_E + T_R \quad (1.10)$$

as the sum of two kinetic energies cannot be negative. In order to initiate such a reaction we need to have a relative energy between P and T. Usually T is kept at rest and P is energised by a device called an accelerator. T is called the target and P is called the projectile. However, if Q is positive, the reaction is energetically possible without any relative energy between P and T. In practice, this process will have a very low probability, as at zero energy the two nuclei may not come close enough to be within the nuclear range if P and T are charged. For P as a neutron, reaction can occur at very low energy. Nucleus P with charge Z_P and nucleus T with charge Z_T will repel each other, and the distance of their closest approach at relative energy E is given by

$$D = \frac{Z_P Z_T e^2}{E} \quad (1.11)$$

Thus the probability of a reaction with positive Q value, though energetically possible, is very small. This is mainly due to the Coulomb barrier between Z_1 and Z_2. If P is a neutron, then there is no Coulomb hindrance, and reaction at very low energy is possible (such as with thermal neutrons of energy 0.25 eV). A nucleus reduces its energy and angular momentum by particle and gamma ray emissions. The relative probability of these emissions depends on several factors. However, in all such emissions, the total energy, angular momentum and parity are conserved quantities.

1.4 EXPERIMENTAL SCENARIO

At the beginning of the 20th century, the structure of the atom was established with the nucleus at the centre and electrons orbiting around it in definite energy states. The next step was to probe the constituents of the nucleus, for which it was necessary to make two or more nuclei interact or react. The simplest process would involve an interaction between two such nuclei, giving rise to what we can call a nuclear reaction. In order to initiate such a reaction, some amount of relative energy was required. As such, historically, alpha particles emitted from radioactive decays were used to carry out the first nuclear reactions. Later, with the invention of accelerators in the 1930s, a large number of reactions could be studied but with light charged particles. The neutrons were also used as one of the interacting nuclei with the advantage that only the nuclear interaction bypassing the Coulomb interaction could be studied. In the 1970s, heavy ion accelerators were discovered, and interactions between heavier nuclei could be studied. The difference between heavier nuclei is that a large Coulomb barrier and angular momentum are encountered. Until the 1990s, stable nuclei were studied. After that nuclei with short half-lives ($\tau_{\frac{1}{2}}$ m s) were produced as radioactive ion beams. The reactions with radioactive ion beams helped in the investigation of very short-lived nuclei that are otherwise not naturally available. The energy in stars is derived from thermonuclear reactions that occur at very low energies. Most of the younger stars synthesise their energy by nuclear reactions involving light charged particles, and so at low energies the probability of these reaction to occur is extremely small. Such reactions were measured and studied from the 1970s with stable and after 1990s with unstable nuclei.

CHAPTER 2

Cross-Section

2.1 EXPERIMENTAL AND THEORETICAL CROSS-SECTIONS

Possibility of a nuclear reaction to occur is measured by a quantity known as the cross-section. As the name signifies, it is the area of the target that one projectile sees during an interaction. Thus its dimension is that of geometrical area. Simply, if the target is considered a circular disk, then the cross-section is πr^2. However, nuclei are very small objects and exhibit probabilistic interaction, so the effective cross-section will be much smaller than the geometrical area. The cross-section is a measurable quantity and will be defined for a practical situation of a nuclear interaction where many projectiles hit a target material that contains many target nuclei. Thus, a normalization has to be done so that the cross-section is defined for a single projectile and a single nucleus. Cross-section of a reaction A(a,b) B is defined as the ratio of the rate (R_b) of a reaction to the incident flux (F_a) and is expressed as

$$\sigma = \frac{R_b}{F_a} \tag{2.1}$$

$$\sigma = \frac{\dfrac{Y_b}{N_A t}}{\dfrac{N_a}{st}} \tag{2.2}$$

DOI: 10.1201/9781003083863-2

where Y_b is the number of nuclei b observed due to interaction with N_A target nuclei of type A over a period of time t; N_A and N_a are the total number of target and projectile nuclei, respectively, and s is the total area of the target nucleus. The cross-section as seen from Equation 2.2 is independent of dynamic parameters such as the number of projectile and target nuclei and time. Instead it is dependent on the static parameters like mass, charge and energy (and sometimes on the angle). In the above expression (Equation 2.2) Y_b is a measurable quantity using nuclear detectors (Chapter 8). The number N_A can be determined if the areal thickness of the target $\left(\mu = \frac{x}{\rho}\right)$ is known, where x is the thickness and ρ is the density of the target material. As A_T (molecular weight) grams contain A_v nuclei (Avogadra number), then

$$N_A = \frac{A_v s \mu}{A_T} \tag{2.3}$$

where $s\mu$ is the mass of the target. In most cases (except for a neutron or photon projectile), the projectile is a charged atom (ion), and

$$N_a = \frac{Q}{ze} \tag{2.4}$$

where ze is the charge state of the beam, i.e. the number of electrons higher or lower than the atomic number. Inserting the values of N_A and N_a, the expression for the total cross-section can be written as

$$\sigma = \frac{Y_b A_T Ze}{A_v \mu t \left(\frac{Q}{t}\right)} \tag{2.5}$$

Inserting the dimensions, with A_T in mass units and μ in mass per area units, Y_b, A_v dimensionless, and Ze and Q having charge units, the cross-section has the unit of area. The last term within the brackets is often called the beam current, $I = \frac{Q}{t}$, and can be measured using a Faraday cup. The definition of a cross-section pertains to the total yield measured over a time t and over the entire solid angle 4π defined with respect to the target centre. Practically it is very difficult to measure the yield over the entire 4π or even a large fraction of it, and so partial cross-sections measured for a small fraction of 4π are more useful.

Theoretically, the cross-section is described in the framework of quantum mechanics, where the rate R_b is determined from Fermi's golden rule as

$$R_b = \frac{2\pi}{h^2} |<\psi_f|V|\psi_i>|^2 \rho_f \tag{2.6}$$

In the above equation, h^2 has the dimension of MeV^2s^2, and $|<\psi_f|V|\psi_i>|^2$ has the dimension of MeV^2. This is because from the normalization condition of probability, i.e.

$$\langle\psi_f|\psi_i\rangle = \int \psi_f^* \psi_i d^3r = 1 \tag{2.7}$$

the dimension of the product of ψ_f^* with ψ_i is fm^{-3}, so that it dimensionally cancels with d^3r, as the right-hand side is unity (no dimension). The term ρ_f is the level density of the available phase space, which is defined as

$$\rho = \frac{d^3rd^3p}{h^3} \tag{2.8}$$

$$= fm^3\left(\frac{MeV}{c}\right)^3 (MeV.s)^{-3} \tag{2.9}$$

$$= MeV^{-1} \tag{2.10}$$

In the above expression, c is the velocity of light in a vacuum. The incident flux is calculated in the framework of quantum mechanics as

$$F_a = \frac{h}{2mi}\left(\psi_{inc}^* \nabla \psi_{inc} - \nabla \psi_{inc} \psi_{inc}^*\right) \tag{2.11}$$

The dimension of the factor outside the bracket is fm^2s^{-1} $\left(\frac{h[MeV.s]}{2m\left[\frac{MeV}{c^2}\right]}\right)$ and that within the bracket is fm^{-4}. Thus the dimension of F_a is $fm^{-2}s^{-1}$.

2.2 PARTIAL AND TOTAL CROSS-SECTIONS

Measurement of the total cross-section over a 4π solid angle is impossible or difficult. Instead, partial cross-sections can be measured over a small solid angle $d\Omega$, i.e.

$$\frac{\Delta\sigma}{\Delta\Omega} = \frac{\sigma}{\Omega} \tag{2.12}$$

where $d\Omega$ is the solid angle subtended by a detector at the target centre. So if the detector of area A is at a distance d from the target, then the solid angle is defined as

$$\Omega = \frac{A}{d^2} \tag{2.13}$$

$\Delta\sigma$ signifies a fraction of the total cross-section σ in a small solid angle $\Delta\Omega$, which is a fraction of the total solid angle, i.e. 4π. The unit used for σ is barn (1 barn = 10^{-27} cm^2), and that of a solid angle is steradian, and therefore the unit of partial cross-section measured over a solid angle is barn.steradian^{-1}. However, in some reactions the detected particle energy may vary over a large range at the angle of detection. In such situation the evaluation of a partial cross-section stands valid for unit energy interval, and its dimension becomes barn.MeV^{-1}.steradian^{-1}. Such a situation happens in compound nuclear and breakup reactions in contrast to scattering and transfer reactions where the emitted particle at a specific angle has a specific energy. This is directly understood from the next chapter on kinematics.

2.3 INCLUSIVE AND EXCLUSIVE CROSS-SECTIONS

If only the emitted particle or radiation of interest is detected without reference to the other products in the reaction, then it is an inclusive measurement. If, on the other hand, all products of a reaction event are detected, then it is an exclusive measurement. As an example, the following reactions are considered:

(a) $^{6}\text{Li}+^{12}\text{C} \rightarrow d+\alpha+^{12}\text{C}$

(b) $^{6}\text{Li}+^{12}\text{C} \rightarrow d+^{16}\text{O}$

The reactions are completely different, but if one detects the deuteron, it is an inclusive measurement and will involve contribution from both reactions. If deuteron is detected in coincidence with ^{16}O, it is an exclusive measurement of reaction (b). Similarly, if deuteron is detected in a triple coincidence with α and ^{12}C, then it is an exclusive measurement of reaction (a). Exclusive measurements are more difficult than inclusive measurements.

2.4 SYSTEM OF REFERENCES-LABORATORY AND CENTRE OF MASS

There are two frames of reference for a nuclear reaction process that are commonly understood. These are known as the laboratory and centre of mass frames. The former is relevant for the experimentalist and the latter for the theorist. In a two-body nuclear reaction $A(a,b)B$, the laboratory frame is the reference from where the observer sees the target nucleus A at rest. This is the practical frame, i.e. where experiments are carried out (targets are at rest in scattering chambers), and if we denote velocity in a laboratory frame with a small letter and that in the centre of mass frame with capital letter, then

$$v_A = 0 \tag{2.14}$$

The projectile velocity v_a is known from the kinetic energy $k_a = \frac{1}{2}m_a v_a^2$, which is generated by a machine known as the accelerator (Chapter 6). The centre of mass of the projectile and target is defined as

$$R_c = \frac{m_a r_a + m_A r_A}{m_a + m_A} \tag{2.15}$$

Where R_c denotes the position vector of the centre of mass with regard to an arbitrary origin and r_a, r_A denote the position vectors of the projectile and target with regard to the existing origin. Differentiating this equation with regard to time, the velocity in the laboratory frame is

$$v_c = \frac{m_a v_a + m_A v_A}{m_a + m_A} = \frac{m_a v_a}{m_a + m_A} \tag{2.16}$$

This equation suggests that the centre of mass moves along the same direction as the projectile and carries a fraction of the magnitude of the projectile velocity. In the centre of mass frame, the observer sees the centre of mass at rest, i.e.

$$V_c = 0 = \frac{m_a V_a + m_A V_A}{m_a + m_A} \tag{2.17}$$

$$m_a V_a = -m_A V_A p_a = -p_A \tag{2.18}$$

Thus in the centre of mass system, the momenta of the two nuclei (either a, A or b, B) are equal in magnitude but oppositely directed. So these incoming or outgoing pairs always lie along a straight line. It is often convenient to use the centre of mass frame in the theoretical calculations, as will be evident in later chapters. However, the centre of mass velocities must have relation with the known quantity, i.e. v_a. This can be established as the two reference frames have relative velocity v_c, and so

$$V_a = v_a - v_c \tag{2.19}$$

$$= v_a - \frac{m_a . v_a}{m_a + m_A} \tag{2.20}$$

$$= v_a \left(1 - \frac{m_a}{m_a + m_A} \right) \tag{2.21}$$

$$= \frac{m_A . v_a}{m_a + m_A} \tag{2.22}$$

Similarly,

$$V_A = v_A - v_c \tag{2.23}$$

$$= 0 - v_c \tag{2.24}$$

$$= -\frac{m_a . v_a}{m_a + m_A} \tag{2.25}$$

The cross-section in the cm and the lab frame are connected by a Jacobian, and this is important because the cross-section is a measurable quantity. Measured cross-sections are in the laboratory frame, and theoretical cross-sections are usually calculated in the cm frame. This is done because in the cm frame, due to equal and opposite momenta, the equations are simplified. The equation of transformation for double differential cross-section from cm to lab is

$$\frac{d^2\sigma}{dV d\Omega} = J\left(\frac{V, \Omega}{v, \omega} \right) \frac{d^2\sigma}{dv d\omega} \tag{2.26}$$

where the capital letters are for the cm system and the lowercase letters are for the lab system quantities. The Jacobian J in the above equation is expressed as

$$J = \begin{vmatrix} \frac{\partial V}{\partial v} & \frac{\partial V}{\partial (cos\theta)} & \frac{\partial V}{\partial \phi} \\ \frac{\partial (cos\Theta)}{\partial v} & \frac{\partial (cos\Theta)}{\partial (cos\theta)} & \frac{\partial (cos\Theta)}{\partial \phi} \\ \frac{\partial \Phi}{\partial v} & \frac{\partial \Phi}{\partial (cos\theta)} & \frac{\partial \Phi}{\partial \phi} \end{vmatrix} \tag{2.27}$$

The x, y and z components of the vector relation between the cm and lab velocity, i.e.

$$V = v - v_c \tag{2.28}$$

are

$$Vsin\Theta cos\Phi = vsin\theta cos\phi \tag{2.29}$$

$$Vsin\Theta sin\Phi = vsin\theta sin\phi \tag{2.30}$$

$$Vcos\Theta = vcos\theta - v_c \tag{2.31}$$

as v_c has only a z component. To obtain the components of the determinant, the equations are differentiated with respect to v, $cos\theta$, ϕ, yielding a set of three equations for the three variables. For example, differentiating with respect to ϕ yields three equations with three variables of the third column in the determinant.

Differentiating the x component equation with respect to ϕ,

$$\frac{\partial V}{\partial \phi} sin\Theta cos\Phi + Vcos\Phi \frac{\partial (sin\Theta)}{\partial \phi} + Vsin\Theta \frac{\partial (cos\Phi)}{\partial \phi} \tag{2.32}$$

$$= vsin\theta \frac{\partial (cos\phi)}{\partial \phi} \tag{2.33}$$

or

$$\frac{\partial V}{\partial \phi} sin\Theta cos\Phi - Vcos\Phi cot\Theta \frac{\partial (cos\Theta)}{\partial \phi} - Vsin\Theta sin\Phi \frac{\partial (\Phi)}{\partial \phi} \tag{2.34}$$

$$= -vsin\theta sin\phi \tag{2.35}$$

Differentiating the y component equation with respect to ϕ,

$$\frac{\partial V}{\partial \phi} sin\Theta sin\Phi + V sin\Phi \frac{\partial (sin\Theta)}{\partial \phi} + V sin\Theta \frac{\partial (sin\Phi)}{\partial \phi} \tag{2.36}$$

$$= v sin\theta \frac{\partial (sin\phi)}{\partial \phi} \tag{2.37}$$

or

$$\frac{\partial V}{\partial \phi} sin\Theta sin\Phi - V sin\Phi cot\Theta \frac{\partial (cos\Theta)}{\partial \phi} + V sin\Theta cos\Phi \frac{\partial \Phi}{\partial \phi} \tag{2.38}$$

$$= v sin\theta cos\phi \tag{2.39}$$

and differentiating the z component,

$$\frac{\partial V}{\partial \phi} cos\Theta + V \frac{\partial (cos\Theta)}{\partial \phi} = 0 \tag{2.40}$$

These three sets of equations with three unknowns, i.e. $\frac{\partial V}{\partial \phi}, \frac{\partial (cos\Theta)}{\partial \phi}, \frac{\partial \Phi}{\partial \phi}$, can be solved to obtain the elements in the third column of the Jacobian. In the same way differentiating with respect to v and $(cos\theta)$ yields, respectively, the elements in columns 1 and 2.

CHAPTER 3

Kinematics

3.1 TWO-BODY KINEMATICS

Kinematics provides the knowledge of energy of the reaction products as function of the angle of the detected particle. This is done by considering the conservation of total energy and linear momentum. This is a classical treatment and is very useful for planning nuclear reaction experiments. In two-body kinematics, nuclear reactions have only two nuclei in the final state, and are considered such as $A(a, b)B$, and from conservation of energy and momenta in the laboratory frame,

$$k_a + Q = k_b + k_B \tag{3.1}$$

$$k_B = \frac{p_B^2}{2m_B} \tag{3.2}$$

$$p_B = 2m_B\left(k_a + Q - k_b\right) \tag{3.3}$$

Considering p_a along the z-direction (i.e. $\theta = 0$ degree),

$$p_B^2 = p_a^2 + p_b^2 - 2p_a p_b cos\theta \tag{3.4}$$

DOI: 10.1201/9781003083863-3

Comparing expressions of p_B in the above equations, a quadratic equation in terms of p_b is obtained:

$$ap_b^2 + bp_b + c = 0 \tag{3.5}$$

where

$$a = \frac{(m_b + m_B)}{m_b} \tag{3.6}$$

$$b = -2p_a cos\theta \tag{3.7}$$

$$c = p_a^2 - 2m_B(k_a + Q) \tag{3.8}$$

Solving the quadratic, the detected particle momentum is

$$p_b = \frac{-b \pm \sqrt{b^2 - 4ac}}{2a} \tag{3.9}$$

Inserting the values of a, b, c from equations

$$p_b = \frac{2p_a cos\theta \pm \sqrt{4p_a^2 cos^2\theta - \frac{4(m_b + m_B)}{m_b}\left(p_a^2 - 2m_B(k_a + Q)\right)}}{2\frac{(m_b + m_B)}{m_b}} \tag{3.10}$$

$$= \frac{p_a \mu_{bB} cos\theta}{m_B} \pm \mu_{bB}\sqrt{\frac{p_a^2 cos^2\theta}{m_B^2} - \frac{2}{\mu_{bB}}\left(\frac{p_a^2}{2m_B} - (k_a + Q)\right)} \tag{3.11}$$

$$p_b^2 = s_1 + s_2 \pm s_3 \tag{3.12}$$

$$s_1 = \frac{p_a^2 \mu_{bB}^2 cos^2\theta}{m_B^2} \tag{3.13}$$

$$s_2 = \mu_{bB}^2\left[\frac{p_a^2 cos^2\theta}{m_B^2} - \frac{2}{\mu_{bB}}\left(\frac{p_a^2}{2m_B} - (k_a + Q)\right)\right] \tag{3.14}$$

$$s_3 = \frac{2p_a\mu_{bB}^2 cos\theta}{m_B}\sqrt{\frac{p_a^2 cos^2\theta}{m_B^2} - \frac{2}{\mu_{bB}}\left(\frac{p_a^2}{2m_B} - (k_a + Q)\right)} \tag{3.15}$$

where $\mu_{bB} = \frac{m_b m_B}{(m_b + m_B)}$ and energy k_b after some simplification is

$$k_b = \frac{p_b^2}{2m_b} = t_1 + t_2 \pm t_3 \tag{3.16}$$

$$t_1 = \frac{m_a}{m_b}\left(\frac{\mu_{bB}}{m_B}\right)^2 k_a cos^2\theta \tag{3.17}$$

$$t_2 = \frac{\mu_{bB}^2}{2m_b}\left[\frac{2m_a k_a cos^2\theta}{m_B^2} - \frac{2}{\mu_{bB}}\left(\frac{m_a k_a}{m_B} - (k_a + Q)\right)\right] \tag{3.18}$$

$$t_3 = \frac{\sqrt{2m_a k_a}\,\mu_{bB}^2 cos\theta}{m_b m_B}\sqrt{\frac{2m_a k_a cos^2\theta}{m_B^2} - \frac{2}{\mu_{bB}}\left(\frac{m_a k_a}{m_b} - (k_a + Q)\right)} \tag{3.19}$$

In the case of elastic scattering, the above equations take a simplified form with $Q = 0$, $m_b = m_a$, and $m_B = m_A$. The elastically scattered energy at angle θ is now represented as k_a', i.e.

$$k_a' = k_a\left[\frac{1}{m_a} - \frac{m_a}{m_A^2} sin^2\theta\right] \tag{3.20}$$

As the energy has to be positive, i.e. $k_a' \geq 0$, from the above equation,

$$sin^2\theta < \frac{m_A^2}{m_a^2} \tag{3.21}$$

$$sin\theta < \frac{m_A}{m_a} \tag{3.22}$$

The above equation is satisfied for all scattering angles for normal kinematics, i.e. when $m_A > m_a$ as $sin\theta < 1$. But for inverse kinematics ($m_A < m_a$) the scattering angle has a maximum value

$$\theta_{\max} = sin^{-1}\left(\frac{m_A}{m_a}\right) \tag{3.23}$$

3.2 THREE-BODY KINEMATICS

Reactions with three nuclei in the final state obey three-body kinematics. Considering such a reaction,

$$a + A \rightarrow 1 + 2 + 3 \tag{3.24}$$

Conservation of total energy and momentum yields

$$m_a + k_a + m_A + k_A = m_1 + k_1 + m_2 + k_2 + m_3 + k_3 \tag{3.25}$$

$$k_a + Q_3 = k_1 + k_2 + k_3 \tag{3.26}$$

$$Q_3 = m_a + m_A - m_1 + m_2 + m_3 \tag{3.27}$$

$$p_a + p_A = p_1 + p_2 + p_3 \tag{3.28}$$

or (with $p_A = 0$) in the laboratory frame,

$$p_3 = p_a - p_1 - p_2 \tag{3.29}$$

and

$$p_3 . p_3 = p_a . p_a + (p_1 + p_2).(p_1 + p_2) - 2p_a.(p_1 + p_2) \tag{3.30}$$

$$p_3^2 = p_a^2 + (p_1^2 + p_2^2 + 2(p_1 . p_2) - 2p_a.(p_1 + p_2) \tag{3.31}$$

$$= p_a^2 + p_1^2 + p_2^2 + 2p_1 p_2 cos\gamma - 2p_a p_1 cos\theta_1 - 2p_a p_2 cos\theta_2 \tag{3.32}$$

where γ is the angle between the momenta p_1 and p_2 with (θ_1, ϕ_1) and (θ_2, ϕ_2) the angle coordinates,

$$cos\gamma = cos\theta_1 \, cos\theta_2 + sin\theta_1 \, sin\theta_2 \, cos(\phi_1 - \phi_2) \tag{3.33}$$

The energy of the third particle is then given as

$$k_3 = \frac{p_3^2}{2m_3} = q_1 + q_2 + q_3 \tag{3.34}$$

$$q_1 = \frac{m_a}{m_3} k_a + \frac{m_1}{m_3} k_1 + \frac{m_2}{m_3} k_2 \tag{3.35}$$

$$q_2 = \frac{2}{m_3} cos\gamma \sqrt{m_1 m_2 k_1 k_2} \tag{3.36}$$

$$q_3 = \frac{2}{m_3} \sqrt{m_a m_1 k_1 k_2}\, cos\theta_1 - \frac{2}{m_3} \sqrt{m_a m_2 k_a k_2}\, cos\theta_2 \tag{3.37}$$

Substituting k_3 from Equation (3.2) in Equation (3.3),

$$\begin{aligned} Q_3 + k_a\left(1 - \frac{m_a}{m_3}\right) = k_1\left(1 + \frac{m_1}{m_3}\right) + k_2\left(1 + \frac{m_2}{m_3}\right) \\ + \frac{2}{m_3} cos\gamma \sqrt{m_1 m_2 k_1 k_2} + \frac{2}{m_3} \sqrt{m_a m_1 k_1 k_2}\, cos\theta_1 \\ - \frac{2}{m_3} \sqrt{m_a m_2 k_a k_2}\, cos\theta_2 \end{aligned} \tag{3.38}$$

3.3 APPLICATION OF KINEMATICS FOR SAMPLE REACTION

The two-body and three-body kinematics calculations are shown for $^{19}F(p,\alpha)$ and $^{19}F(d,n\alpha)$ reaction. The variation of alpha particle energy with respect to its emission angle is shown in Figure 3.1. This calculation helps in choosing the type of detector and its thickness in nuclear reaction experiments. As the energy is much greater at forward angles compared to that at backward angles, a much thicker detector material is required at forward angles.

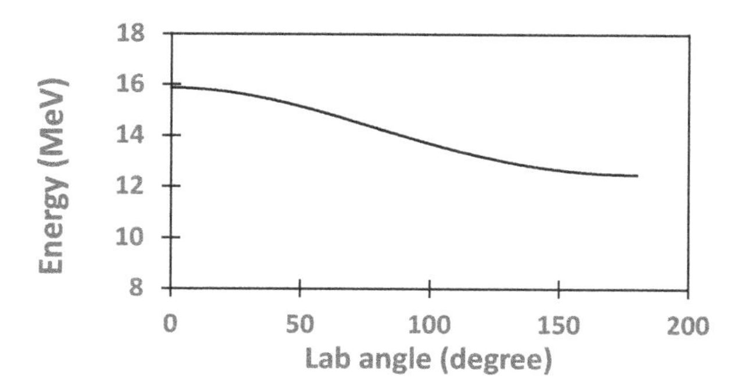

FIGURE 3.1 Variation of α particle energy with respect to emission angle in $^{19}F(p,\alpha)$ reaction.

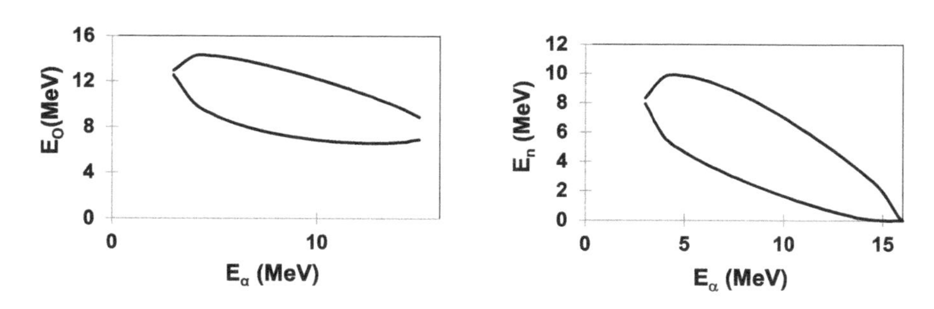

FIGURE 3.2 Variation of energy of any two of the three particles emitted in $d+^{19}F$ reaction.

The application of three-body kinematics is shown in Figure 3.2 where the energies of any two outgoing particles are plotted in the reaction $d+^{19}F \rightarrow \alpha+^{16}O+n$.

CHAPTER 4

Types of Nuclear Reactions

4.1 CLASSIFICATION OF REACTIONS

Reactions can be broadly classified depending upon time scales of interaction or by the number of products they produce in the final state. There are a number of reactions that occur in a timescale which is required for the projectile to traverse the nuclear dimension. Since the nuclear dimension is of the order of fermi and energy is of the order of MeV, the timescale can roughly calculated as

$$t = \frac{d}{v} \tag{4.1}$$

$$= \frac{d}{\sqrt{2E/m}} \tag{4.2}$$

$$t = \frac{d}{c\sqrt{2E/mc^2}} \tag{4.3}$$

where d is the nuclear dimension, E is the energy of projectile, mc^2 is the mass in MeV and c is the velocity of light. For a proton of mass 931.8 MeV the time calculated from the above equation using 1 fermi as the nuclear dimension and E as 1 MeV is 3.3×10^{-22} s. This is the typical time scale of a

DOI: 10.1201/9781003083863-4

nuclear reaction if the interaction involves only a few interactions before the final products are emitted. Much slower interactions are also observed which involve a large number of interactions before the final products are produced. These timescales are of the order of 10^{-14} s and involving electromagnetic emissions. Direct and compound reactions represent these two extreme timescales. Reactions have been also classified into two-body and three-body final state forming reactions. They show widely different characteristics in terms of energy spectra and angular distributions. Transfer reactions that produce two-body final states have discrete energy states, whereas breakup reactions, a three-body reaction, show a continuous energy spectrum.[6,7] The discovery and production of short-lived neutron- or proton-rich nuclei has led to a plethora of studies on such breakup reactions, as extra nucleons added to stable nuclei to form radioactive nuclei have very small separation energies. As such, they easily break in the field of another nucleus. Reactions with heavy ions have quite different behaviour in comparison to light ions. Greater amounts of angular momentum can be transferred in reactions involving heavier nuclei. This is because angular momentum is proportional to the reduced mass of the two nuclei, $L = \mu vr$. Sometimes nuclear reactions are also classified in terms of the incident energy. Usually for reactions with incident energies well below the Coulomb barrier ($= \frac{Z1Z2e^2}{R}$, where R is the touching radius of P and T) are known as sub-Coulomb energy reactions. Such reactions are pertinent to thermonuclear reactions in stars. Reactions above the Coulomb barrier are highly probable and form the crux of classical nuclear physics. Energies above the Fermi Energy (33 MeV per nucleon) are known as intermediate energy reactions.

On the basis of timescales, nuclear reactions can be broadly classified as direct and compound reactions at the two extremes. Direct reactions can result in both two-body and three-body final states. The one which results in two final states can again be of two types. One is when the initial pair of nuclei P and T do not change to a different set of two nuclei E and R. Such a process is known as scattering. Scattering may be of two types: elastic and inelastic. In elastic scattering the initial pair of nuclei remain the same after the nuclei interact. Only the kinetic energy and the direction of motion are altered. In inelastic scattering, one or both the initial nuclei get excited to one of its low-lying states after the interaction. In two-body reactions, apart from scattering, there may be rearrangement or transfer reactions. In this process a new set of nuclei are produced. A part of the projectile gets transferred to the target, giving rise to a stripping (transfer) reaction. On the other hand, if the projectile picks up a part of the target due to an interaction, then a pickup (transfer) reaction occurs. In either elastic or inelastic scattering and transfer reaction the Q values are either zero or very small. So they are sometimes collectively known as quasielastic (like elastic) reactions.

On the other hand, there may be a massive transfer reaction in which a large nuclei or a large number of nucleons may be transferred, and in this scenario a large Q value is observed. These are known as deep inelastic reaction or massive transfer reactions. Three-body reactions such as breakup or knockout also form a part of the direct reaction family. In a breakup reaction the projectile fragments into its constituents in the field of the target nucleus. These reactions, however, become favourable when the projectile nucleus has a loosely bound structure. Stable loosely bound nuclei ^{6,7,9} and most of the neutron-rich radioactive nuclei are candidates for projectile breakup reactions. If, on the other hand, a light nucleus or nucleon knockouts represent a constituent of the target producing a three-body final state, a knockout or target breakup process occurs.

The compound nucleus reaction is a two-body reaction where the final products are formed by the decay of an intermediate nucleus formed by the amalgamation of the projectile and target. This amalgamation is known as fusion reaction. A fusion leads to the formation of an excited compound nucleus that mainly decays by emission of light particles, namely neutrons, protons, deuteron, helions, tritons, and alphas. This is known as evaporation reaction and represents the major decay mode of the compound nucleus. The emission of these light particles de-excites the nucleus. If it does not possess any more energy, it decays by gamma emission. These γ-rays only decay between the low-lying states of the compound nucleus and hence bear the signature of their structure. Statistical gamma rays can also be emitted competing with particle emissions. The compound nucleus can also decay by massive two fragment emissions known as fusion–fission reaction. However fission occurs mainly in the actinide region ($Z>92$). In some lighter nuclei asymmetric fission occurs resulting in orbiting reaction.

Intermediate between the direct and compound reaction is a reaction that is known as a pre-compound reaction. In this process the particles are emitted after a few interactions more than in direct reactions but before the formation of the compound nucleus. Precompound reactions are usually observed at 10 MeV per nucleon energies. They are often associated with doorway states, i.e. states that may lead to either direct reactions or to a compound nuclear reaction.

4.2 SCATTERING

In the scattering process, as mentioned earlier, there can be two types of scattering: elastic and inelastic. Elastic scattering process can be expressed as

$$ {}_{Z1}^{A1}P + {}_{Z2}^{A2}T \rightarrow {}_{Z1}^{A1}P + {}_{Z2}^{A2}T \tag{4.4} $$

where the final products are the same as the initial pair of nuclei. From the definition of Q, the value given in equation in elastic scattering $Q = 0$ and from equation kinetic energy is conserved. Example of an elastic scattering process is ^{6}Li + ^{12}C → ^{6}Li+^{12}C or $^{12}C(^{6}Li,^{12}C)^{6}Li$.

In inelastic scattering either the projectile or target or both are excited to a higher energy level. This can be expressed as

$$ {}_{Z1}^{A1}P + {}_{Z2}^{A2}T \rightarrow {}_{Z1}^{A1}P^{*} + {}_{Z2}^{A2}T \tag{4.5} $$

$$ {}_{Z1}^{A1}P + {}_{Z2}^{A2}T \rightarrow {}_{Z1}^{A1}P^{*} + {}_{Z2}^{A2}T^{*} \tag{4.6} $$

$$ {}_{Z1}^{A1}P + {}_{Z2}^{A2}T \rightarrow {}_{Z1}^{A1}P + {}_{Z2}^{A2}T^{*} \tag{4.7} $$

Conservation of energy for inelastic reaction of a sample projectile excitation can be written as

$$ M_P c^2 + T_P + M_T c^2 + T_T = M_P c^2 + T_P' + E^* + M_T c^2 + T_T' \tag{4.8} $$

where E^* is the excitation energy of the projectile P. Therefore,

$$ Q = \left(T_P' + T_T' - T_P + T_T \right) = -E^* \tag{4.9} $$

Example of inelastic scattering is the excitation of ^{12}C to its first excited state at 4.42 MeV (2^+), i.e. ^{6}Li + ^{12}C → ^{6}Li + $^{12}C^{2+}$. Elastic scattering spectrum results in a single peak-like structure as shown in Figure 4.1. The energy of this peak can be predicted by the kinematics. In inelastic scattering several excited states may be also seen in Figure 4.1 and the energies are less than the elastic peak energy.

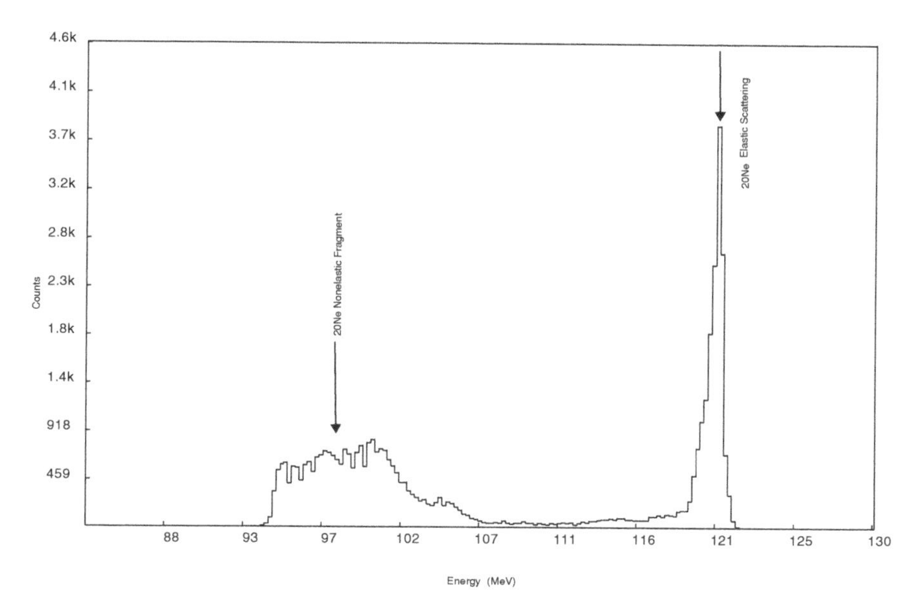

FIGURE 4.1 Elastic and inelastic scattering cross-section.

4.3 DIRECT REACTION

Direct reactions generally involve a few interactions between the projectile and target nucleons. As such, only a small fraction of momentum is transferred onto the target. The particles emitted from direct reactions have a forward peaked angular distributions, i.e. most of the particles are emitted mostly at small angles with respect to the incident direction. A direct reaction can be expressed as

$$ {}_{Z1}^{A1}P + {}_{Z2}^{A2}T \rightarrow {}_{Z3}^{A3}E + {}_{Z4}^{A4}R \tag{4.10} $$

where E and R are, respectively, the ejectile and the residual nuclei. So the difference from scattering is that the incident pair of particles will change in the case of a reaction. Transfer reaction is an example of a two-body directs reaction. If we consider that P in Equation (4.1) has a two-component structure, i.e. $P = Q + R$, then either Q or R can be transferred to the target T giving rise to E i.e.

$$ P(=Q+R)+T \rightarrow E(=T+Q)+R \tag{4.11} $$

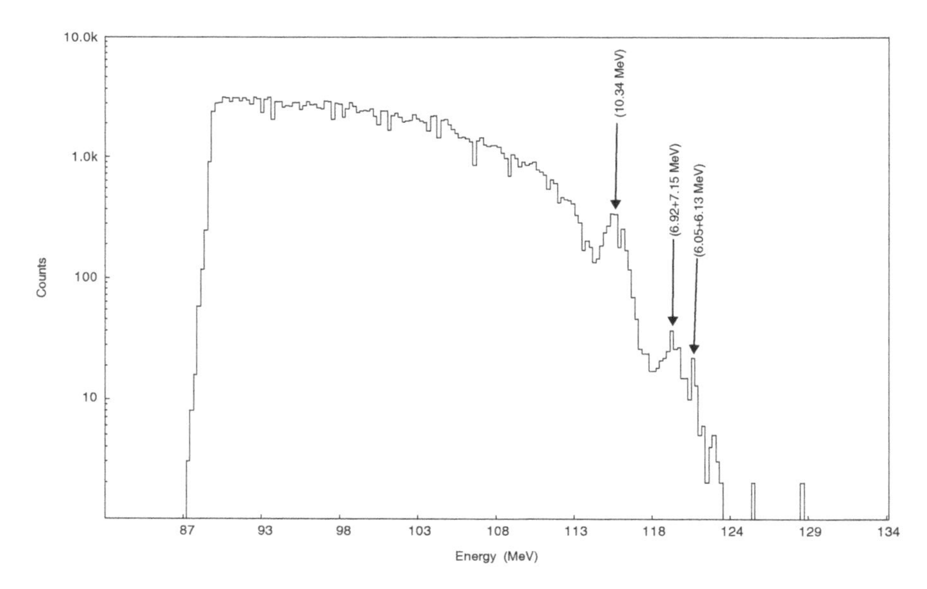

FIGURE 4.2 Discrete states populated in a nuclear reaction.

An example of a transfer reaction (α-transfer) is ^{6}Li + ^{12}C → d + ^{16}O. If one measures the cross-section of E at various angles of its emission, then a large fraction of the total cross-section will be in the forward cone. This is because of the very few interaction and small momentum transfer between P and T. In transfer reaction the energy spectrum is a number of discrete peaks as shown in Figure 4.2. The peaks correspond to the ground and excited states of the residual nucleus R. Thus one can perform particle spectroscopy (study of various energy levels) of a nucleus using transfer reaction. Besides the two-body reactions, there are three-body direct reactions as well. They are known as breakup reactions, where Q is very loosely bound to the nucleus P (i.e. the separation energy of Q or R from P is small), i.e.

$$P(=Q+R)+T \rightarrow Q+R+T \tag{4.12}$$

The two stable nuclei ^{6}Li and ^{7}Li have a very low alpha breakup threshold (1.46 and 2.47 MeV, respectively) and exhibit large breakup cross-sections. An example of a breakup (projectile breakup) reaction is ^{6}Li + ^{12}C → α + d + ^{12}C. In a knockout reaction, as the name suggest, a nucleon can knock out a component from the target nucleus that has a two-component structure, i.e.

$$P + T(= T1 + T2) \rightarrow P + T1 + T2 \tag{4.13}$$

In both breakup and knockout reactions, the energy spectrum, unlike in direct reactions and scattering, is continuous in nature (see Figure 4.4) due to energy being shared among more than two particles (similar to the continuous beta decay spectrum that led to the proposal of a third particle, the neutrino). The maximum cross-section of a breakup occurs at an energy where the fragments of the projectile (Q or R) move with beam velocity. Let the beam velocity be v_P, then energy of the breakup fragment Q is given by

$$E_Q = \frac{m_Q v_P^2}{2} = \frac{m_Q c^2}{2} \frac{v_P^2}{c^2} \tag{4.14}$$

where

$$\frac{v_P}{c} = \sqrt{\frac{2E_P}{m_P c^2}} \tag{4.15}$$

So that

$$E_Q = \frac{m_Q c^2}{2} \frac{2E_P}{m_P c^2} = \frac{m_Q}{m_P} E_P \tag{4.16}$$

Similar expression for the other fragment can be obtained.

4.4 COMPOUND AND PRECOMPOUND REACTIONS

Compound nuclear reactions stand at the other extreme of the direct reaction mechanism. In this reaction process the final nuclei are produced through an intermediate nucleus. A compound reaction can be represented as

$$^{A1}_{Z1}P + ^{A2}_{Z2}Q \rightarrow ^{A}_{Z}C \rightarrow ^{A3}_{Z3}E + ^{A4}_{Z4}R \tag{4.17}$$

The conservation law demands

$$A_1 + A_2 = A = A_3 + A_4 \tag{4.18}$$

$$Z_1 + Z_2 = Z = Z_3 + Z_4 \tag{4.19}$$

The angular momentum conservation is same as for direct reactions except that the total angular momentum before and after the reaction must be equal to the total angular momentum and parity of the compound nucleus as well, i.e.

$$\vec{l}+\vec{S}=\vec{J}=\vec{l}'+\vec{S}' \tag{4.20}$$

$$(-1)^{l}\Pi_P\Pi_T=\Pi_C=(-1)^{l'}\Pi_E\Pi_R \tag{4.21}$$

The first part of a compound reaction, i.e.

$${}^{A1}_{Z1}P+{}^{A2}_{Z2}Q\rightarrow{}^{A}_{Z}C \tag{4.22}$$

is known as fusion or the formation. The second part, i.e.

$$C\rightarrow{}^{A3}_{Z3}E+{}^{A4}_{Z4}R \tag{4.23}$$

is known as evaporation if E or R is a nucleon (neutron or proton) or a light nuclei or cluster (such a α, triton, helion, deuteron) and fission if the E and R are both heavy nuclei. ^{6}Li + ^{12}C → ^{18}F* → n + ^{17}F and n + ^{235}U → Kr + Sr+ xn are examples of evaporation and fission reactions. With reference to the B/A curve, a fusion will release energy if C is lighter than nickel. This is because if we choose two nuclides lighter than nickel and fuse them, and if they form a compound nucleus that is lighter than nickel, energy will be released. This is understood from the B/A curve. The Q value for fusion reaction will be

$$Q=(M_P(A_1,Z_1)+M_T(A_2,Z_2)-M_C(A,Z)c^2 \tag{4.24}$$

Replacing the mass in terms of respective binding energies,

$$Q=Z_1M_p+(A_1-Z_1)M_n-B(Z_1,A_1) \tag{4.25}$$

$$+Z_2M_p+(A_2-Z_2)M_n-B(Z_2,A_2) \tag{4.26}$$

$$-(ZM_p+(A-Z)M_n-B(Z,A) \tag{4.27}$$

$$=B(A,Z)-B(Z_1,A_1)-B(Z_2,A_2) \tag{4.28}$$

as $A = A_1 + A_2$ and $Z = Z_1 + Z_2$. As a fraction of the mass number of the compound nucleus A, the Q value is

$$\frac{Q}{A} = \frac{B(Z,A)}{A} - \frac{B(Z_1,A_1)}{A} - \frac{B(Z_2,A_2)}{A} \tag{4.29}$$

$$= \frac{B(A,Z)}{A} - \frac{A_1}{A}\frac{B(Z_1,A_1)}{A_1} - \frac{A_2}{A}\frac{B(Z_2,A_2)}{A_2} \tag{4.30}$$

With both A_1 and A_2 less than A and $\frac{B(Z_1,A_1)}{A_1}$ and $\frac{B(Z_2,A_2)}{A_2}$ much smaller than $\frac{B(Z,A)}{A}$ (see Figure 1.3 and observing that binding energy per nucleon increases rapidly with the small change in mass number), $\frac{Q}{A}$ is positive in most cases and energy will be released in regions where $A < 60$. The opposite happens in the region $A > 60$. There energy is released if the compound nucleus splits into two nuclei instead of fusing two nuclei. The same argument used earlier can be used with expression for Q value reversing for the decay scenario. The splitting phenomena is known as fission and is more likely in the region where $Z > 92$. This because in such heavy nuclei many protons are far apart and have a weaker binding (due to short-range nuclear force) compared to the Coulomb repulsion. Fission fragments are neutron rich (as lighter fragments require fewer neutrons than protons to become stable compared to their mother) and decay by emitting neutrons. The fission process is dynamic and evolves from the point of a highly deformed nucleus (saddle) to a splitting nucleus (scission). An interesting question is which nuclei will be fissile, or why nuclei with $Z > 90$ are more fissile. The test can be done easily by a calculation of the difference of the Q value and the Coulomb barrier (V^c) for fission, i.e.

$$Q - V_c = (M(A_{F1}, Z_{F1}) + M(A_{F2}, Z_{F2}) - M(A_C, Z_C) \tag{4.31}$$

$$= -\frac{Z_{F1}Z_{F2}e^2}{R_1 + R_2} \tag{4.32}$$

The difference is known as activation energy, and if it is positive, then fission can occur spontaneously. Energy spectra of compound nuclear

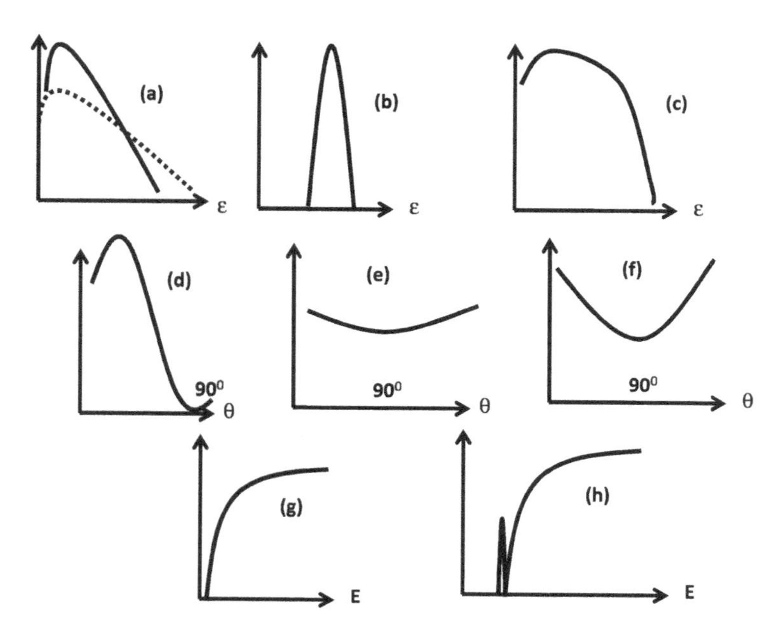

FIGURE 4.3 Energy spectra for (a) compound nucleus and precompound (b) direct (c) breakup reaction; angular distribution in (c) direct (d) and (e) compound emission; excitation function for (e) non-resonant and (f) resonant reaction.

emissions show a Maxwellian shape. Both evaporation particles and fission fragments are emitted either isotropically (equally probable at all angles) or with a symmetry around 90 degrees in the centre of the mass frame. This is in contrast to direct reactions where the probability maximizes at the forward angles. Precompound reactions show features in between direct and compound reactions. These reactions are generally observed at incident energies lower than at which direct reactions occur and at energies higher than where compound reactions dominate. The energy spectrum in a precompound process show a Maxwellian shape as for the evaporation particles but with two distant slopes (Figure 4.3a) corresponding to two temperatures (one higher and one lower). Excitation functions, energy spectra, and angular distributions for different reactions are shown in Figure 4.3.

4.5 ASTROPHYSICAL REACTIONS

Stars in the universe synthesize elements through nuclear reactions, thereby generating enormous amounts of heat and light. The Sun is a star that is relatively young and small. Sizes of all other stars are expressed in terms of the Sun's mass. The larger the star, the larger its gravitational self-energy (as gravitational force depends on the masses of the two objects they are acting on), as the centre of a massive body experiences the maximum gravitational force (as it is surrounded by the maximum matter). As a result, a star begins contracting and eventually would collapse on itself quite quickly if there were no elements in it providing the balancing expansion force. Due to contraction, gravitational energy decreases, and the released energy is converted to thermal energy or heat. The temperature of the star increases and can vary from a few million Kelvin (10^6 K) to hundreds of Giga Kelvin (10^9 K). This temperature in the MeV scale can be obtained by the relation

$$E = kT \tag{4.33}$$

where k, the Boltzmann constant, is $8.617 \times 10^{-11-11}$ MeV/K. As such, except for the hottest stars, the available energy for a nuclear reaction is very small. In a newborn star the most abundant element is hydrogen, and so two hydrogen nuclei can react and produce energy (fusion) as described in Section 4.4. However the two nuclei are charged and experience a Coulomb repulsion. The fusion cross-section depends upon the tunnelling probability of one of the interacting nuclei with respect to the other through this barrier. As the relative energy is much below the Coulomb barrier, the tunnelling probability is very small. The same is true for other light elements. This energy released in fusion goes partially to oppose the gravitational contraction and partially released as heat and light. The energy release from fusion counteracts the gravitational collapse, but now the temperature decreases as energy from the decrease in gravitational energy has ceased. Thus hydrogen fusion also ceases and temperature decreases. But again gravitational collapse starts, and if the star is large, higher temperatures (or higher relative energy) are reached and nuclei heavier than hydrogen can fuse, such as helium, carbon, silicon etc. Charged particle fusion or burning occurs till the A = 60 mass region is reached. Beyond this region, as emphasised earlier, fusion does not release energy. So how are the heavier elements synthesised now? Actually during the light ion fusion reactions, some intermediate reactions generate neutrons. $^{13}C(\alpha,n)$ and $^{22}Ne(\alpha,n)$ are examples of such reactions. These neutrons can be captured to form neutron-rich nuclei beyond A = 60 mass region (e.g. $^{58}Ni(n,\gamma)^{59}Ni$. However as more neutron-rich isotopes are produced by this process, the nuclei so produced are more and more unstable to beta decay and thereby form stable nuclei. So one can compare the rate of neutron capture to beta decay. If the beta decay rate is faster than the neutron capture rate, then more stable nuclei are produced. This is known as the rapid capture or the *r*-process nucleosynthesis. The opposite scenario produces more unstable nuclei, the process known as slow or *s*-process. In the *s*-process, more neutron-rich nuclei are produced. In this way the whole range of nuclei along the stability line and on the neutron-rich side are produced. The rate of neutron capture increases as a function of neutron density inside the star. The proton-rich or neutron-deficient nuclei are produced by proton capture but their formation cross-section is lower due to the Coulomb barrier. Some nuclei have very low proton capture cross-sections, and these nuclei are known as "waiting point" nuclei. The proton capture process compete with positron decay or electron capture gives rise to either proton-rich or stable nuclei. Thus the formation of proton-rich nuclei is known as *rp*-process or rapid proton capture process. Since at low energies the most dominant reaction mechanism is compound nuclear reaction, most of the reactions occurring in the nucleosynthesis process are compound nuclear in nature. Moreover, the exit channel for low-energy reaction is an electromagnetic emission (gamma emission) except for the elastic channels or some (p,α) or (α,p) channels that have positive *Q*-values. The fusion cross-section of any two nuclei is of utter importance to the scientists. As stars of different sizes provide different relative energy or temperature, the fusion cross-sections of two nuclei at different energies are required. This variation of fusion cross-sections with relative energies of the two reacting nuclei is known as the fusion excitation function. The excitation function at a particular relative energy is one of the most challenging measurements in nuclear astrophysics. The cross-section at which this measurement is required is known as the Gamow energy. The definition of Gamow energy comes from the requirement of calculation of astrophysical reaction rate. As these reactions are driven by the temperature of the star, they are also known as thermonuclear reactions. The rate of the reaction at an energy *E* is defined as

$$\langle \sigma v \rangle = \int_0^\infty \sigma(v) M(v,T) dv \tag{4.34}$$

where $\sigma(v)$ is the fusion cross-section at relative velocity v ($v = \sqrt{\frac{2E}{m}}$) and $M(v, T)$ is the Maxwellian distribution function for a particle at velocity v and temperature T. The Maxwellian distribution is given by the standard expression

$$(v,T) = \left(\frac{m}{2\pi kT}\right)^{1/2} e^{-\frac{mv^2}{2kT}} dv \tag{4.35}$$

The fusion cross-section $\sigma(v)$ increases with increasing velocity, and $M(v, T)$ decreases with increasing velocity. A plot of the two functions that constitute the integrand of the integral in the above equation is presented in Figure 4.4. The common region is enveloped by the two functions and has a maximum energy known as the Gamow energy (E_G). The reaction rates have to be determined at Gamow energy, and for a charged particle–induced reactions, this energy is much smaller than the Coulomb barrier. For a particular pair of nuclei the Gamow energy changes with the change in temperature. This is understandable, as the Maxwellian distribution is dependent on temperature, and so the convolution part should also be dependent on temperature.

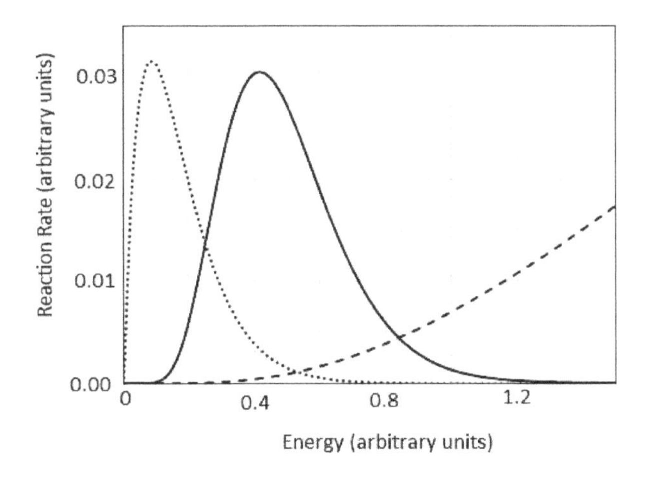

FIGURE 4.4 Gamow peak (solid line), fusion excitation function (dashed line), and Maxwell–Boltzmann energy distribution (dotted line).

Measurement of fusion cross-sections of various reactions in the stars at the Gamow energy is the crux of experimental nuclear astrophysics. However, the reactions involving nuclei of mass $A < 60$ are charged particle–induced reactions. As such, at Gamow energy, the cross-section is very low. Measurement of such low cross-section with good accuracy is very difficult. The reason is that with low cross-section, a large amount of time is required to accumulate a large count (see Section 2.1). This is required to have a small statistical uncertainty (see Section 10.1). In order to achieve good statistical accuracy, a thick target or high beam current is required. Simultaneous use of both these may generate a huge amount of heat in the target, thereby damaging it unless special efforts are adopted. Counting over long time for improving the statistics is often a technical challenge, as the electronic instrumentation may suffer from stability problems. Nuclear reactions at low energy often decay by gamma emission only. Measurement of low gamma cross-sections is hampered by cosmic and room background effects. Therefore, shielding of detectors, vetoing the background and underground measurements are necessary. All these pose additional difficulties in measurement of fusion cross-sections at very low energies. The typical variation of fusion cross-section with the relative energy is shown in Figure 4.3. This graph is also known as fusion excitation function and indicates many things. The fusion cross-section decreases with decreasing energy but starts decreasing drastically below a certain energy level. This energy is equal to the Coulomb barrier between the interacting charged particles. The cross-section below the Coulomb barrier decreases drastically as the tunnelling probability decreases. In a classical picture there would have been zero probability below the Coulomb barrier, but in a quantum scenario there is always a finite probability for the particle to penetrate the barrier. As the energy is lower than the Coulomb barrier, the particle has a finite probability of tunnelling through it, and this decreases as the energy of the particle decreases. But once it tunnels through the barrier, it gets trapped in the attractive nuclear potential, and fusion is possible. The tunnelling probability ($T(E)$) through the Coulomb barrier can be parameterised as

$$T(E) = e^{-2\pi\eta} \tag{4.36}$$

where η is the Somerfield parameter and is defined in terms of the relative velocity v as

$$\eta = \frac{2Z_1Z_2e^2}{\hbar v} \tag{4.37}$$

Since $v = \sqrt{2E/m}$, the dependence of the tunnelling probability on E is clearly understood. However, the cross-section has dependence on the nuclear interaction, and therefore it is parameterised as

$$\sigma(E) = \frac{T(E)S(E)}{E} \tag{4.38}$$

where $S(E)$ is known as the astrophysical S-factor and contains the nuclear contribution in the tunnelling process. There is significance in this form of parameterisation of the cross-section. As has been emphasised earlier, $\sigma(E_G)$ cannot be determined with sufficient accuracy for most reactions. Therefore, an extrapolation is done using the cross-section at energies greater than the Gamow energy. As the cross-section is changing very rapidly with energy, the error in the extrapolation is likely to be very large. This rapidly varying nature mainly comes from the Coulomb term and is well known. The nuclear term is very slowly varying but is the unknown part in the cross-section. This unknown, slowly varying nuclear part is factorised as $S(E)$. Thus extrapolation of $S(E)$ becomes more fruitful than extrapolation of $\sigma(E)$ except for when there is a resonance (a sharply varying cross-section at a particular energy compared to its neighbouring energies). However, in the above discussion, the relative orbital angular momentum is neglected. This is because at low energy, the relative angular momentum l is almost zero. This can be understood from the expression for $l = m(r \times v)$ where we can see that dependence on E comes through v. The orbital angular momentum presents a repulsive centrifugal barrier to the sum of nuclear and Coulomb potential that makes the nuclear potential shallower (see Section 11.2). At higher energy, fusion reaction involves higher l values, and so the increase of fusion probability becomes more gradual than at lower energies. Finally at a very large angular momentum, the attractive nuclear potential becomes so shallow that fusion becomes impossible and fission occurs. This angular momentum is known as critical angular momentum. Sometimes the cross-section of astrophysical reaction is extracted from a different reaction than the actual reaction. This is called the indirect method. The indirect method uses a different nuclear reaction with a higher cross-section than the actual astrophysical reaction to determine. The Asymptotic Normalization Constant (ANC), Trojan Horse, Coulomb dissociation and resonance elastic scattering are some of the indirect methods used in nuclear astrophysics.

CHAPTER 5

Nuclear Reaction Experiments

5.1 BASIC FACILITIES REQUIRED FOR EXPERIMENTAL NUCLEAR REACTION STUDIES

It is important to know what the basic infrastructure facilities are required to study nuclear reactions. Some of these facilities are expensive and time consuming to setup. In nuclear reaction studies the detectors can be quite complex and may vary from experiment to experiment. The purpose of this chapter is to introduce the facilities, their basic principle of operation, approximate cost that may be involved and make the reader aware how to plan some basic experiments using these basic facilities.

The first and foremost facility required to perform a nuclear reaction in the laboratory is an accelerator. An accelerator (see Chapter 6) is a device that energizes a particle. Most of the accelerators deal with charged atoms or ions but there are accelerators for neutrons and gamma rays as well. To understand the principle of an accelerator, we consider an electron drifting in a potential difference V. The energy gained by the electron in travelling through the potential difference is $E = eV$. This is the principle of a DC accelerator. Of course, higher the value of the potential difference, the higher is the energy obtained. As seen in Section 1.2, to initiate negative Q value reactions, an energy of the order of million electron volt will be required. So for a nuclear reaction to occur, a million-volt potential difference will be required. This is a very high voltage, and its stability is also

DOI: 10.1201/9781003083863-5

important to have a constant value of energy. The principle of operation of different accelerators will be discussed in Chapter 6.

Another essential facility required for nuclear reaction experiment is a vacuum chamber. A vacuum chamber is an enclosure (usually made of stainless steel or aluminium) that can be evacuated to very low air pressure. This pressure for nuclear reaction experiments can be about 10^{-6} millibar whereas atmospheric pressure is 1,000 millibar. The vacuum chamber can be of different size and shape depending upon the requirement, i.e. depending upon the radiation to be detected. If charged particles are to be detected, then a large scattering chamber is required. One of the reasons for this is that charged particle detectors need to be placed inside a vacuum to avoid unwanted or complete energy loss in the air. There are other reasons, which are discussed in the next subsection. On the other hand, if gamma rays or neutrons are detected in the experiment, then a small vacuum chamber is sufficient to house the target. The detectors can be kept in the air, as gamma rays or neutrons will have minimal interaction with air before reaching the detectors. In order to increase the geometric efficiency of the detectors (see Chapter 8) and detect different particles emitted from the same nuclear event, multi-detector arrays are used. In these arrays sometimes charged particle detectors are combined with neutron and gamma detectors. The path leading to the charged particle detector should be in a vacuum. A simple schematic array is shown in Figure 5.1 to clarify which regions need be in a vacuum for proper operation. However, whether a scattering chamber or an array is required depends upon the criteria for the experiment. Besides using arrays, some experiments use separators or spectrometers. These are devices that use magnetic field in their detection process. It is well known that a charged particle moves in a circular path under the influence of a magnetic field, and the radius of this orbit depends upon the charge-to-mass ratio and the magnetic field. Choosing different magnetic fields will generate different radii for particles with different q/m. Thus a cocktail stream of particles having different q/m are separated, and the beam of interest is focussed by some electromagnetic lens (see Section 6.5) to a point called the focal point. At this point the particles are again identified by a detector (usually known as a focal point detector). This detector uses particle identification method (based upon Bethe Bloch relation, see Section 8.6) to separate charges and masses in a beam of fixed q/m. For example, a deuteron and alpha particle will have the same $q/m = 2$ and so will be present in the

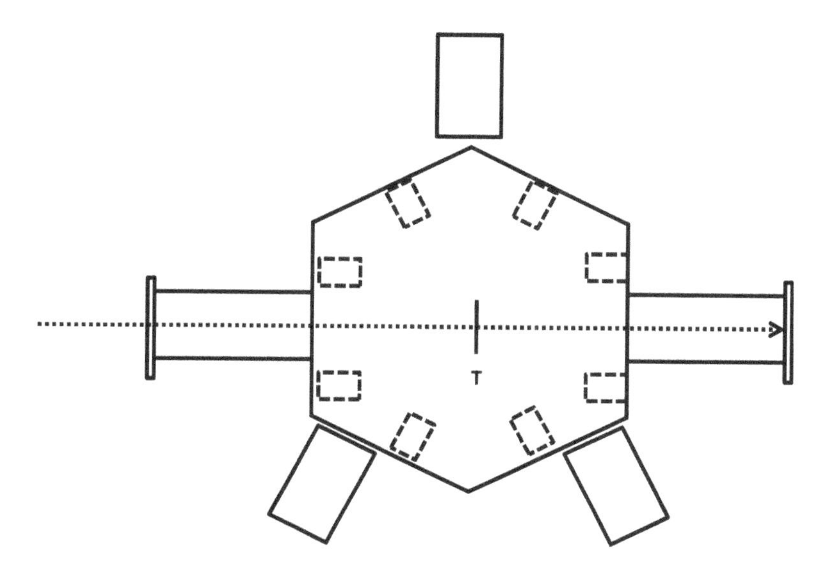

FIGURE 5.1 A multi-detector array.

same focal point. But since they have different masses and charges, they can be separated by a charged particle detector telescope that acts as a particle identifier. This may be a ΔE-E type detector or a single detector like the Bragg Curve detector.

5.2 LARGE SCATTERING CHAMBER EXPERIMENT

A large scattering chamber is an important facility for nuclear reaction experiments. Typically a large chamber has a diameter of 1–2 m. The chamber can have a wide range of facilities. A schematic of a large scattering chamber is shown in Figure 5.2. The question is why we need such a large facility. One reason has been discussed before. Another reason is that for nuclear scattering studies we often require cross-section measurement at different angles. This is known as angular distribution and provides very important indication about a reaction. Now if one has a large distance from the detector to the target, then the error in the determination of the angle is reduced. Moreover, in a larger chamber one can accommodate a large number of charged particle detectors. Detectors with gaseous medium are often very bulky and require space, and only a large chamber can house such bulky detectors. A scattering chamber can be of various shapes and sizes depending upon the requirement. The most common type with the associated facilities is discussed with reference to the Figure 5.2.

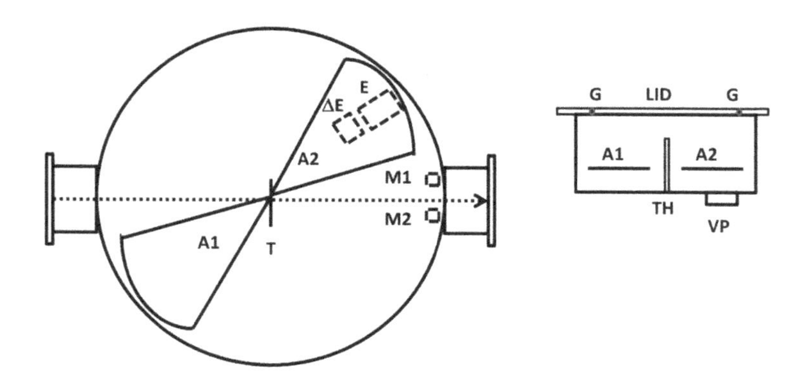

FIGURE 5.2 A large scattering chamber.

The figure shows a cylindrical scattering chamber that has an openable lid which is provided with a vacuum-tight seal by a gasket on the chamber body. The chamber has many ports placed diametrically along the height of the cylinder. These ports are either used for viewing inside the chamber during vacuum or can be used for mounting detectors or some electronic modules. At the centre there is an upright arrangement (T) to house the nuclear targets. This arrangement can be rotated or shifted up and down without disturbing the vacuum from outside. The detector arms (A1, A2) are an important component of the chamber. These arms usually have provision for angular movement in fine angular steps so that few detectors can be rotated at various angles to take data. This avoids the use of a very large number of detectors, which adds immensely to the complication in an experiment. The chamber should be attached to a vacuum pump that can evacuate the chamber to a pressure no worse than 10^{-6} millibar. In order to achieve this vacuum, one needs to have a very clean environment inside the chamber. All materials used other than the metallic parts should be of low vapour pressure, such as the cable sheaths, detector mounting spacers, glues etc. The two ports kept for beam in and beam out should have provision to house collimators to achieve a highly collimated beam onto the target. This facilitates a very good definition of the angle to be measured. Another usefulness of a large scattering chamber is to perform time of flight measurement of charged particle detectors (Figure 5.3) that requires as large a separation as possible between the two detectors D1 and D2. However, often a large chamber can be very expensive and difficult to install, and so many of the facilities have to

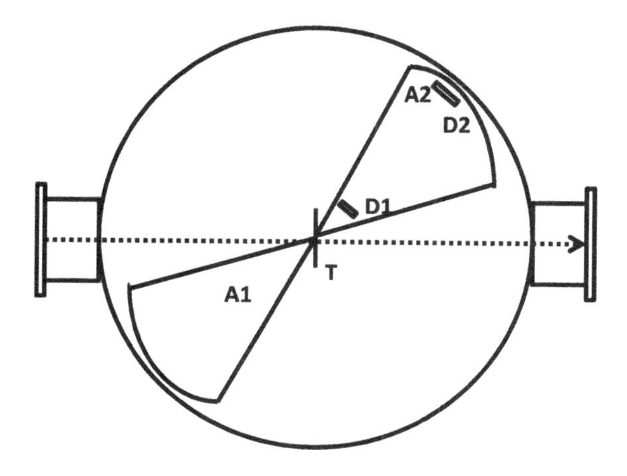

FIGURE 5.3 TOF setup.

compromise. Described below are the two simple scattering chamber experiments that will explain its utility.

EXPERIMENT 1: RUTHERFORD SCATTERING ANGULAR DISTRIBUTION

In this experiment we need a beam of projectile, say α particle, and a thin target of any material. A thin target is required, as in an elastic scattering (Rutherford scattering is an elastic scattering due to the repulsive Coulomb interaction between the nuclei of the projectile and the target atoms) we need to detect the same particle as the projectile. A very thick target will either stop the alpha particle or change its energy substantially. But we only want to know the change in energy of the projectile due to the nuclear scattering process and not due to electronic interaction with the target atoms. In the simplest form of experimental setup we would like to put a detector that can detect charged particles. The best and simplest detector that one can use is a silicon surface barrier detector (Section 8.5). The detector is placed as shown in Figure 5.2 at some angle (θ) with respect to the beam direction. If now the beam falls on the target, interaction occurs and particles emerge in different directions from the target. Depending upon the energy of the projectile and the Q values for different reactions, particles will be emitted. If the projectile α energy is much less than the Coulomb barrier of the projectile-target system (where Coulomb barrier is defined as $\frac{Z_p Z_t e^2}{R_p + R_t}$ with $R_i = r_0 A_i^{1/3}, i = p,t$ and A_i is the mass number of the target or projectile and r_0 = 1.2 fermi), then only Rutherford scattering is dominant. The angle of the detector can be obtained from the circular scale attached with the detector arms and placed outside the vacuum. It is important to select the alpha particle and measure its energy in the experiment. This is because though at low energy Rutherford scattering is dominant, other reaction processes cannot be completely ignored. The selection of the alpha particle from other particles or nuclei can be done by using two silicon detectors instead of one. The two detectors are placed one after the other as shown in Figure 5.2. The front detector is usually thin enough to let the alpha particle penetrate it. We can, of course, calculate the energy of the alpha particle from simple two-body kinematics as described in Section 3.1. The back detector is kept thick enough to stop the alpha in it. A charged particle looses energy in matter by ionisation following the Bethe–Bloch relation as described in Section 8.6. The infinitesimal energy loss dE of a charged particle of mass M, charge state Z (electronic charge not nuclear charge) and energy E in passing through an infinitesimal length dx is from Bethe relation,

$$\frac{dE}{dx} \propto \frac{MZ^2}{E} \tag{5.1}$$

Thus we can write:

$$E.\frac{dE}{dx} = k.MZ^2 \tag{5.2}$$

The last equation is that of a hyperbola (xy = constant), and one hyperbola appears for each set of particles with different M and Z. As we have already ensured that the detector we are using can measure energy, we can plot the range of energies measured by the front detector in the y axis along with the corresponding residual energies measured by the back detector along the x axis. This plot will generate a number of hyperbolae as shown in Figure 5.4. If the total intensity is counted, the experimental cross-section can be obtained by the procedure described in Section 2.1.

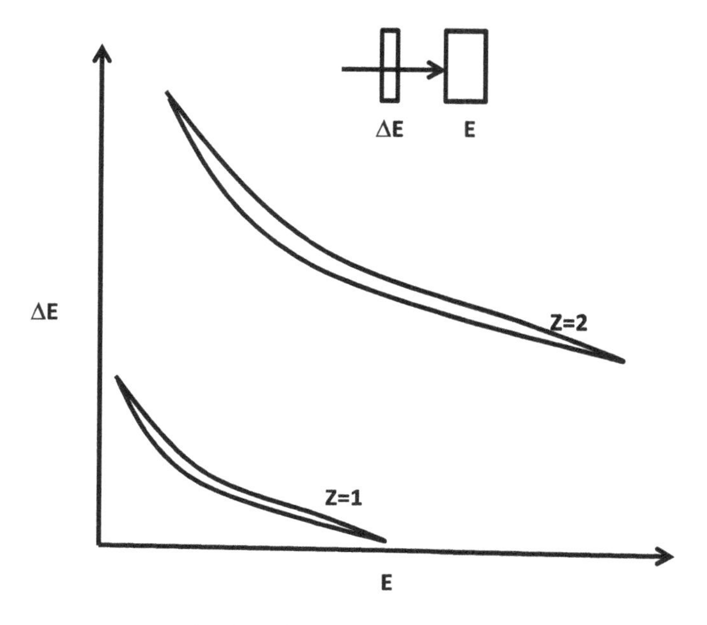

FIGURE 5.4 ΔE-E particle identifier spectrum.

EXPERIMENT 2: TIME OF FLIGHT (TOF) EXPERIMENT

This is another experiment that can be used for determination of velocity of a particle as well as one that can identify various particles using this technique. The TOF setup is shown in Figure 5.3, and as can be seen, this experiment requires two detectors (D1, D2) with good time resolution. This means that these detectors can measure the time of arrival of a particle very accurately. As an example, a particle of velocity v_1 is considered that passes through the first detector D_1 at t_1 and passes detector D_2 at time t_2. If D_1 and D_2 are separated by a length L, then the velocity of the particle can be obtained by a ratio of the length to the time difference (or time of flight), i.e. $\frac{L}{t_2 - t_1}$. Let us consider a proton and alpha particle of same energy 10 MeV passing through the two detectors. As their masses are different, the velocities are also different. Considering masses of proton as 931.5 MeV and that of alpha as 4 × 931.5 MeV, the velocities, respectively, will be 0.1465c and 0.07c, where c is the velocity of light. In order to keep these velocities well resolved, one needs to measure the time difference quite accurately. If the flight path L is very short, then the time difference is compressed and resolving the velocities becomes very difficult. Suppose two flight lengths are chosen as 30 cm and 90 cm. The TOF for the situation is an advantage, as now the experiment is feasible even without a good timing resolution of the two detectors. Thus a larger chamber makes the TOF setup more convenient.

5.3 SMALL CHAMBER EXPERIMENT

For nuclear reaction experiments where only gamma rays and or neutrons are detected, a small vacuum chamber to house the target is enough. This is because gamma rays and neutrons have negligible interaction with air, and so the respective detectors can be kept outside vacuum unlike charged particle detectors. In a small chamber experiment, targets can be made thick enough to stop the beam and measure the beam current onto the target. Sometimes large detector arrays may be attached with a small target chamber instead of using piece wise detectors in a large scattering chamber. An array for charged particles can be made up of a multi-tier proportional counter in front followed by a Bragg curve detector and a thick solid state detector such as a CsI detector. The multi-wire and Bragg detectors are gas medium detectors and are separated from the target chamber by a thin window foil. Large detector arrays provide very high geometrical efficiency.

5.4 EXPERIMENTS WITH SEPARATORS AND SPECTROMETERS

Separators and spectrometers are basically magnetic devices that are used (like purifying the beam from an ion source in an accelerator up to the target) after the beam interacts with a target and a cocktail of particles are emitted (cocktail in the sense that these particles have different charge state to mass ratio and energies or momentum). If we consider a particle of charge q and mass $'m'$ and moving with velocity v in a magnetic field of strength B, then the particle experiences a Lorentz force ($F = \vec{q} \times \vec{B}$) that tends to move the particle perpendicular to the plane containing v and B. The path of the particle is that of a helix. In case the magnetic field is kept perpendicular to the velocity, then the motion follows a circular path of radius ρ. The radius is obtained by equating the Lorentz force with the centrifugal force, i.e.

$$\frac{mv^2}{\rho} = Bqv \tag{5.3}$$

$$B\rho = \frac{mv}{q} \tag{5.4}$$

$B\rho$ is known as the magnetic rigidity and depends on the momentum of the particle and charge. By adjusting the magnetic field the radius of the particle can be made such that the particle passes through the slits placed after the magnet. The particles with different q/m will have a different ρ and will be blocked by the slits. Practically, however, particles having q/m close to the desired particle may have similar radii and may pass through the slits. In order to make a secondary selection a ΔE detector is placed after the spectrometer magnet. However, the selected beam may have a divergence amongst its charges, and so a focussing quadrupole is placed between the spectrometer magnet and the detector. A quadrupole is a four-pole piece with two north and two south poles. A quadrupole will shift an ion towards the axis depending upon the direction of the beam current.

As described in Section 6.5, a quadrupole, if focussed along the x-axis, will defocus in y, and vice versa. So at least two quadrupoles are required to focus the ion in both x and y axes on the detector. A $\Delta E - E$ measurement further separates the desired ion. A spectrometer provides better resolution, as there is a two-course filtering of the beam in terms of M, Z and E.

A separator is a similar magnetic device without the detector. A separator also selects an ion of specific charge q/m from a cocktail beam. This is important when one wants to detect an event that has a much lower cross-section compared to the beam. In the production of radioactive ion beam, an intense stable beam bombards a target from which, due to a suitable reaction, the rare ions are emitted. The cross-section of the rare ions is 7–8 orders of magnitude smaller compared to stable ions. As the beam forms the major background after the target, a separator can clean up the situation considerably.

5.5 EXPERIMENTAL SETUPS FOR ASTROPHYSICAL REACTIONS

As discussed in Section 4.4, the astrophysical reactions pose a great challenge for measurements, as these reactions occur at energies far below the Coulomb barrier of the interacting charged particles (except for neutron-induced reactions). At such low energies in most reactions gamma emission is the only possible reaction channel. As cross-sections are low, the background gamma emission detected poses a great obstacle. So shielding of this background radiation is essential in studying astrophysical reactions in the laboratory. The background radiations can be classified into gamma rays that are either below or above 3 MeV. The gamma rays that are below 3 MeV are mainly of terrestrial origin, i.e. they are coming from Earth itself. The main gamma radiations from terrestrial origin are from ^{40}K and ^{226}Rn nuclei. The other radiations come from nuclei that belong to the ^{232}Th and ^{238}U radioactive decay series. The energy of gamma rays of terrestrial origin is below 3 MeV and can be shielded from gamma detectors by a layer of Pb. The gamma rays of extraterrestrial origin are greater than 3 MeV and originate from cosmic ray–induced reactions. These high-energy gamma rays cannot be stopped by a shielding and therefore pose a great difficulty in detecting gamma rays from astrophysical reactions that one tends to study – more so, as many reactions will emit gamma rays above 3 MeV. One way to reduce the cosmic background is by setting up the detection laboratory underground beneath several meters of rocks. As such, these experiments become very expensive. These are, however, passive shielding methods. An active shielding method is carried out when an extra detector is placed behind the actual detector and gamma rays of higher energy than that of interest penetrate the actual detector and reach the extra detector. A anticoincidence (veto) logic between the two gamma detectors then eliminates the gamma rays that are of higher energy and not of astrophysical interest.

CHAPTER 6

Accelerators for Nuclear Reactions

As evident in the earlier chapter, nuclear reaction study requires a device called the accelerator. An accelerator is a machine that energises a nuclear particle: light ions, heavy ions and even neutrons. A charged particle is accelerated simply by putting it in an electrical potential difference. However, as the energy required in nuclear reaction is of the order of MeV (million electron volt), the potential difference should be of the order of million volts. Such high potential cannot be produced by an ordinary electrical generator, and sustaining such high voltage for a long time without a breakdown requires technological specialties. There are basically two types of nuclear accelerators, cyclic and DC, and they differ in the way they energise the particle.

6.1 BRIEF HISTORY OF ACCELERATORS

The discovery of accelerator brought about a revolutionary change in nuclear physics research. Cyclic and DC accelerators were first invented almost at the same time in 1929. These were the Cyclotron by E.O. Lawrence and the Van de Graaff generator by R.J. Van de Graaff. The first linear accelerator was invented by R. Wideroe at about the same time. The Cockroft–Walton Generator was invented a little later in 1932.

DOI: 10.1201/9781003083863-6

6.2 BASIC DEFINITIONS AND CLASSIFICATION OF ACCELERATORS

An accelerator essentially makes a charged particle jump up a potential difference and thereby force it to gain energy. Thus an electron of charge e, when put in a potential difference of V volts, gains an energy E where

$$E = eV \tag{6.1}$$

In case of a heavy ion, it is qeV where q is the charge state of the ion. The charge state of an ion is the number of electrons the neutral atom has lost from its neutral configuration. For example, a carbon atom that has six electrons in its neutral configuration acquires a charge state of +1. The maximum charge state of a carbon atom is +6, i.e. when it is fully stripped of all the electrons and becomes a bare carbon nucleus. An accelerator is technically a very complex and expensive device. It basically consists of an ion source, an accelerating device, beam-optical and beam diagnostic components. The ion source creates an ion of the desired nucleus to be accelerated by various methods. The main concept is to create a plasma (ionised gas) that has a high density of the ion. The plasma can be either of a positive or negative charge that is accelerated by a comparatively low voltage (kV). The ion is further carried through beam pipes in high vacuum (so that the ion does not lose energy) under an attractive potential and is energised. In this process some accelerators change the charge state of the ion, thereby achieving higher energy. This is called a tandem accelerator, as they increase the energy in steps (usually in two steps). In a tandem accelerator (Figure 6.1), in the ion source usually a single charge state negative ion is created. The single negative ion is extracted with a potential usually of the order of a few tens of kilovolt (e.g. 30 kV). The single charge state negative ion is then attracted at the central region, which is of a very

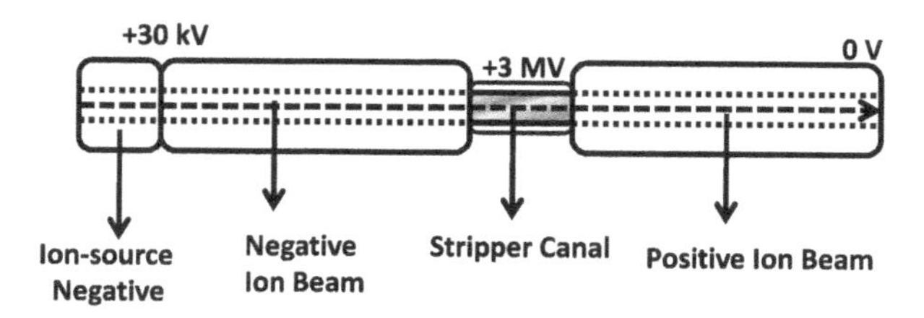

FIGURE 6.1 Tandem accelerator.

high positive potential (of the order of 3 MV). The devices that can generate such high voltage are discussed later in this chapter. At the central region the energy of the ion is extraction energy (EE) plus the acceleration energy, i.e. EE+eTV (TV is the terminal voltage). In the example shown in Figure 6.1, it is 30 keV+3.MeV at the terminal. At the terminal there is a stripper canal where either a thin stripper foil (carbon) or a stripper gas at a fixed low pressure (nitrogen) is kept. As there is a very high vacuum on either side of the stripper canal, differential pumping is carried out so that the gas does not diffuse out on either side. The purpose of the stripper foil or gas is to change the charge state of the ion (except for hydrogen, which has one electron and is already stripped, so the stripper canal has no function). As a result, the ion beam is now positively charged and again undergoes an acceleration towards the right (in Figure 6.1) where the potential is at the ground level (ideally 0). If the charge state acquired by the ion is q, then the total energy gained in the process is $EE + eTV + qeTV = EE + (q + 1)TV$. In the next section the generation of high voltage V by some popular nuclear accelerators will be discussed.

6.3 CHARGED PARTICLE ACCELERATORS

The charged particle accelerator starts with an ion source followed by an accelerating or high voltage generating device. The most common generators are the (a) Van de Graaf or Pelletron and (b) Cockroft–Walton. These are generators for DC acceleration. In the Van de Graaff and Pelletron devices the method of charge induction is used, and a system of charge transport system transfers the charge to a terminal and raises it to a high potential. The charge transport is either by a belt (Van de Graaff) or by a series of pellets (Pelletron). The basic mechanism is shown in Figure 6.2.

Owing to the necessity for charge transport, these accelerators are vulnerable to breakdown if disruption in the belt or pellets occur. The Cockroft–Walton accelerator, on the other hand, does not use any moving system and are more stable. However, much higher energy can be achieved in the Pelletron accelerators.

(i) Van de Graaf and Pelletron Generator

As shown in Figure 6.2, there are two pulleys P1 and P2 which are connected by a belt. The material of P1 is one that accepts electrons when it is rubbed with a neutral material. In this case when the pulleys are brought into rotation by a motor, P1 due to friction with the

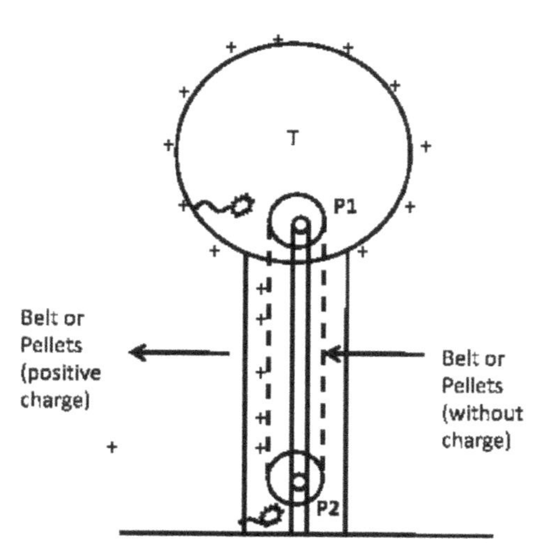

FIGURE 6.2 Van de Graaff and Pelletron accelerators.

belt (which is made of nylon a neutral material) acquires electrons and the belt positive charge. The negative charge on P1 induces a strong positive charge on the comb spikes. The very small surface area helps in accumulating a large amount of positive charge on each spike. The large positive charge on the spike and negative charge on P1 create a strong electric field and ionises the air molecules in the gap between the comb and P1. The positive charges (ions) from this ionisation are attracted towards P1 and, as the belt comes, in its way gets onto it and is transported at the top. The top pulley P2 is made of a material that accepts positive charges when rubbed by a neutral material. The positive charges on P2 induce large negative charge on the top comb spike heads and again an electric field is formed. The electric field ionises the air between P2 and the top comb, and the positive charges are attracted towards the spike. The positive charges on the belt additionally repel positive charges onto the spikes of the top comb and are transferred to the terminal dome where they are uniformly distributed. The terminal is thus profusely charged up to a high potential.

(ii) Cockroft–Walton Generator

The Cockroft–Walton accelerator, on the other hand, does not use any moving system and is more stable as they use semiconductor devices like diodes. This voltage generator uses a voltage multiplier circuit using diodes and capacitors and amplifies the magnitude of the input AC voltage to a DC voltage, with magnitude an integral multiple of the input amplitude. The integral multiple depends upon the number of diodes and capacitors used. In the figure a simple voltage doubler circuit is used where two diodes and two capacitors are used. The input is a sinusoidal AC voltage $v_{in}(t) = v_0 sin\omega t$, and the voltage across C_2 is the output voltage and is $2v_0$. However, this voltage does not appear at the output instantaneously and requires a finite charging time with a large number of cycles (at least 20 to 25). So the circuit will be very slow with a normal household AC supply, and so a radiofrequency power supply is required. The charging mechanism can be understood from the circuit shown in Figure 6.3.

Initially at $\omega t = 0$ the two capacitors C_1 and C_2 have no charge and voltage across them is zero. Therefore, the potentials at A and B are the same and equal to $v_{in}(t)$. The condition for diode D_1 to conduct is $v_{in}(t) < 0$ and that for D_2 to conduct is $v_{in}(t) > 0$. In the cycle between $0 < \omega t < \frac{\pi}{2}$ the condition for D_2 to conduct is satisfied and D_1 becomes open. The equivalent circuit is now the two capacitors in series with the AC input. The capacitors are charged equally to half the maximum voltage v_0 in this cycle, i.e. $\frac{v_0}{2}$. In this cycle the output voltage is $\frac{v_0}{2}$. The voltage at B is now $v_{in}(t) - \frac{v_0}{2}$ using Kirchoffs Law, and condition for D_1 to conduct becomes $v_{in}(t) - \frac{v_0}{2} < 0$ or $v_{in}(t) < \frac{v_0}{2}$, and D_2 to conduct is $v_{in}(t) - \frac{v_0}{2} > \frac{v_0}{2}$ or $v_{in} > v_0$. So D1 conducts in

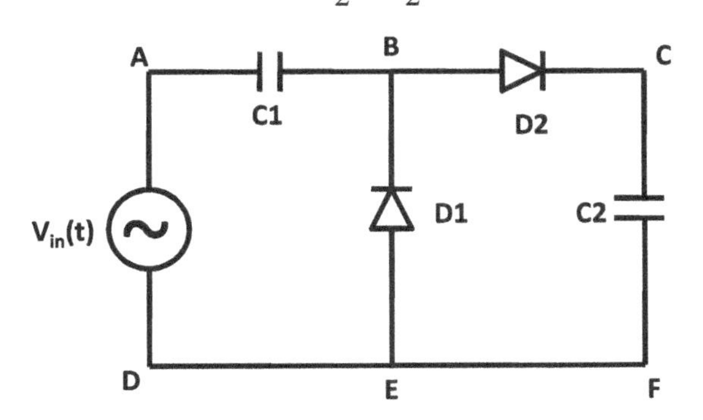

FIGURE 6.3 Cockroft–Walton generator (voltage doubler).

this cycle, and the output voltage does not change. The voltage at B also remains constant at $v_{in} - \frac{v_0}{2}$ until $v_{in}(t) < \frac{v_0}{2}$, after which the capacitor starts charging through D1. As D1 is shorted, $v_B = v_E = 0$, and again by Kirchoff's law $v_{AB} = v_{in}(t)$, and at the end of the cycle at $\omega t = \pi$, $v_{AB} = 0$ and $v_{BE} = v_{int}(t)$. In the cycle $\pi < \omega t < \frac{3\pi}{2}$, condition for D2 to be ON is $v_{in} > \frac{v_0}{2}$, which is not possible in this cycle, as $v_{in}(t)$ is always negative till $\omega t = \frac{3\pi}{2}$. In this cycle D1 is charged to $-v_0$ and voltage at B at the end of the cycle is $v_{in}(t) + v_0$. After this cycle D1 is OFF, as $v_{in} + v_0 < 0$ cannot be satisfied. The condition for D2 to be ON is $v_{in}(t) + v_0 > \frac{v_0}{2}$ or when $v_{in}(t) > -\frac{v_0}{2}$. So in the cycle where $\frac{3\pi}{2} < \omega t < 2\pi$, till $v_{in}(t) < -\frac{v_0}{2}$ the output voltage remains unchanged at $\frac{v_0}{2}$. After the input voltage becomes greater than $-\frac{v_0}{2}$, D2 conducts, and the equivalent circuit is as shown in Figure 6.3 by shorting D2. The total voltage change is $\frac{v_0}{2}$, and change of voltage at the end of the cycle on each capacitor is $\frac{v_0}{4}$. So the output voltage or the voltage across C2 is $\frac{v_0}{2} + \frac{v_0}{4}$, i.e. $\frac{3v_0}{4}$. Voltage on C1 is $-v_0 + \frac{v_0}{4} = -\frac{3v_0}{4}$. The voltage at B in the beginning of the next cycle of the input voltage when $2\pi < \omega t < \frac{5\pi}{2}$ is $v_{in}(t) + \frac{3v_0}{4}$. So the condition for D1 to conduct is $v_{in}(t) + \frac{3v_0}{4} < 0$ and that for D2 to conduct is $v_{in}(t) + \frac{3v_0}{4} > \frac{3v_0}{4}$, i.e. $v_{in}(t)$ should be positive. So up to $\omega t = \frac{5\pi}{2}$ D1 is OFF and D2 is ON. At $\omega t = 3\pi$, the output voltage is calculated again considering the two capacitors in series as before with voltage $-\frac{3v_0}{4}$ on C1 and $\frac{3v_0}{4}$ on C2. The change in input voltage is v_0, and so the output voltage is changed by $\frac{v_0}{2}$ on each capacitor. Output voltage at $\omega t = 3\pi$ will therefore become $\frac{3V_0}{4} + \frac{v_0}{2} = \frac{5v_0}{4}$. It can be now

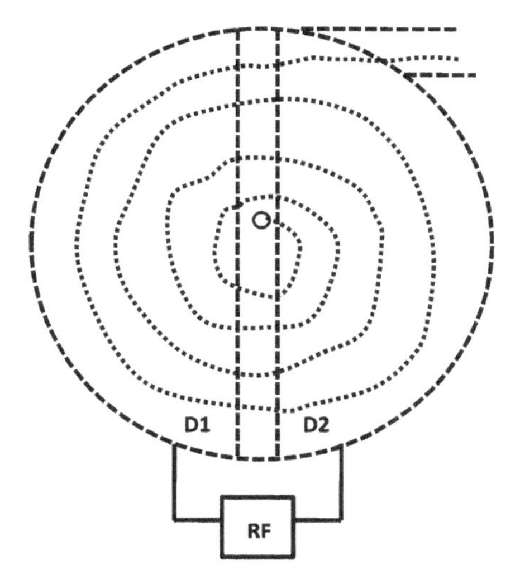

FIGURE 6.4 Cyclotron.

realised that the circuit would take several cycles of the input voltage to raise the output voltage to $2v_0$. So a high-frequency input power supply is preferred for the Cockroft–Walton device.

(iii) Cyclotron

The cyclotron is a cyclic accelerator in the sense that the beam travels in a circular path instead of in a straight line. The basic structure of the cyclotron is shown in Figure 6.4. A radiofrequency power supply is attached across the two D's. The frequency is adjusted to match the rotational velocity of the charged particle in a way that when the particle reaches the gap after each half-cycle, the polarity of the power supply changes so that the ion experiences an attraction and an energy gain. After getting an initial energy from the ion source, a semi-circular path inside the 'D' is due to the Lorentz force exerted by the magnetic field on the particle of charge $q = (eZ)$ and the centrifugal force experienced by the particle, i.e.

$$\hat{F} = q\left(\hat{v} \times \hat{B}\right) = \frac{mv^2}{r} \tag{6.2}$$

$$= qvBsin\theta = \frac{mv^2}{r} \tag{6.3}$$

$$= qvB = \frac{mv^2}{r} \tag{6.4}$$

In a cyclotron, as evident from the figure, the motion of the particle and the magnetic field are in planes perpendicular to each other, and so the angle $\theta = 90°$. The rotational frequency from equation is

$$\omega = \frac{v}{r} = \frac{qB}{m} \tag{6.5}$$

This is known as the cyclotron frequency, and the time period of the path is

$$T = \frac{2\pi}{\omega} \tag{6.6}$$

which is independent of the radius and energy of the particle, but depends only on the mass and charge state of the ion. In an electron beam, owing to its lighter rest mass than nuclei, relativistic mass variation has to be considered, which is again dependent on velocity. The maximum energy that can be attained by a particle of mass m and q in a cyclotron of magnetic field B and radius R of the 'D' is

$$mv = qBR \tag{6.7}$$

$$E = \frac{m^2v^2}{2m} \tag{6.8}$$

$$= \frac{q^2B^2R^2}{2m} \tag{6.9}$$

or the energy per nucleon is

$$\frac{E}{A} = \frac{q^2B^2R^2}{2m_0A^2} \tag{6.10}$$

$$= K\left(\frac{q}{A}\right)^2 \tag{6.11}$$

where

$$K = \frac{(BR)^2}{2m_0} \tag{6.12}$$

In the above equation, BR is the magnetic rigidity, m_0 is the mass of the proton and A is the mass number of the nucleus. The K of the cylotron is a distinctive parameter giving an indication of the energy that can be delivered by it. Examples are K130 Cyclotron of VECC, India or K500 Cyclotron of MSU, USA. It also is interesting to note that, at a point when the radius of the particle is r, the equations of motion are

$$mv = qBr \tag{6.13}$$

$$r = \frac{mv}{qB} \tag{6.14}$$

Differentiating the last equation, the variation in r is

$$\Delta r = \frac{m\Delta v}{qB} \tag{6.15}$$

So the change in r decreases with increase in energy.

6.4 NEUTRON GENERATORS

Neutrons cannot be energised, as a charged particle and a nuclear reaction is used to generate neutrons of specific energy. Thermal neutron of very low energy (0.25 eV) can induce fission in actinide nuclei ($Z > 92$). These neutrons are produced in nuclear reactors and help in propagating a chain reaction. But thermal neutrons are not suitable to initiate large negative Q value reactions. The 14 MeV neutron generator is most commonly used for such purpose. The $t(d,\mathrm{n})\alpha$ reaction is used, which has a Q value of 17.589 MeV. An RF ion source using deuterium gas is used for producing a low-energy deuteron beam of few 100 keV. Using conservation of energy for the $t(d,n)\alpha$ reaction in the centre of mass frame,

$$E_{cm} = K_n + K_\alpha \tag{6.16}$$

$$E_{cm} = \frac{m_t}{m_p + m_t} E_{lab} \tag{6.17}$$

$$= \frac{3}{5} \times 0.1 \tag{6.18}$$

$$= K_n + K_{alpha} \tag{6.19}$$

$$= K_n \left(1 + \frac{1}{4}\right) \tag{6.20}$$

The centre of mass energy of the neutron from the above relations comes out to be 14.0712 MeV. Hence the name: 14.1 MeV neutron generator. The neutron beam has high radiation hazard problems and is difficult to collimate.

6.5 ION OPTICS AND BEAM DIAGNOSTIC DEVICES

Ion beams, unlike the neutrons, can be diagnosed and controlled using ion optical elements that are basically electric and magnetic fields. When an ion beam passes through the beam line in a vacuum towards the chamber, housing the target material, its dimension gets larger due to repelling action of the ions amongst themselves. This is known as defocussing of the beam, which is an unwanted feature. Defocussing of the beam results in loss of beam current, which in turn results in lower yield and therefore would require more time for experiment to measure the cross-section in comparison to when the beam current is higher. In scattering experiments, beam dimensions should be as small as possible to reduce the errors in the scattering angle of the beam particles after interaction at the target. Theoretically the beam is considered to be a point particle, so focussing of the ion beam is important. This is brought about by a quadrupole (an arrangement of four magnetic poles), which may be magnetic or electric. The focussing action of a quadrupole is explained in Figure 6.5. The quadrupole has a focussing action in Y and defocussing in X. The force on

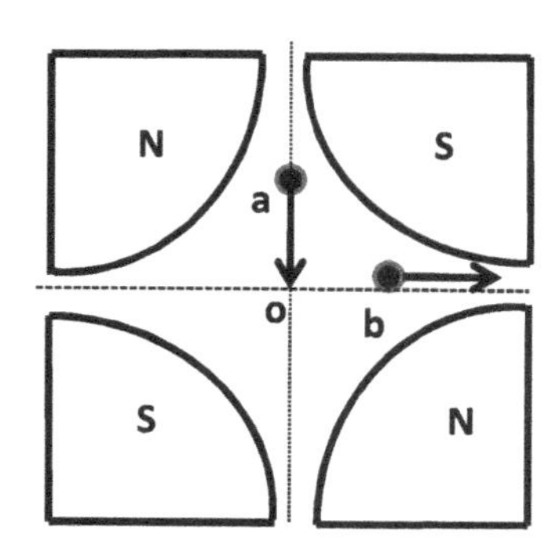

FIGURE 6.5 Quadrupole.

the ion is obtained by using Flemmings left-hand rule, where the current direction is the direction of motion of the positive ion, field direction is from north to south pole, and motion determines whether the beam is pushed away or towards the beam axis. The quadrupole lens is more suitable at higher energy levels.

If the beam is of very low energy, then focussing can be brought about by an Einzel lens whose action can be understood from Figure 6.6. Besides the quadrupole, magnets constitute essential ingredient of beam optical elements. Magnets are able to bend the ions depending upon their mass and charge state. This is utilised both in bending and switching magnets used in an accelerator. The bending or switching magnet uses a dipole magnet (Figure 6.7) to bend the beam, and the selection of desired species is possible by cutting off the other species by a slit.

The switching magnet can also select and mainly divert the beam in different directions so that the user can set up their experimental stations separately and the beam can be switched depending upon the requirement. Several devices are used for monitoring and diagnosing the beam. The Faraday cup (FC) and the beam profile monitors (BPM) are the most useful devices for respectively measuring the beam current and observing the beam cross-section in the xy plane.

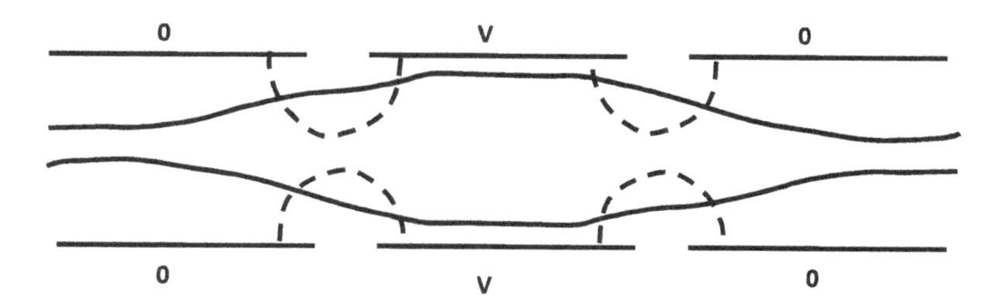

FIGURE 6.6 Einzel lens.

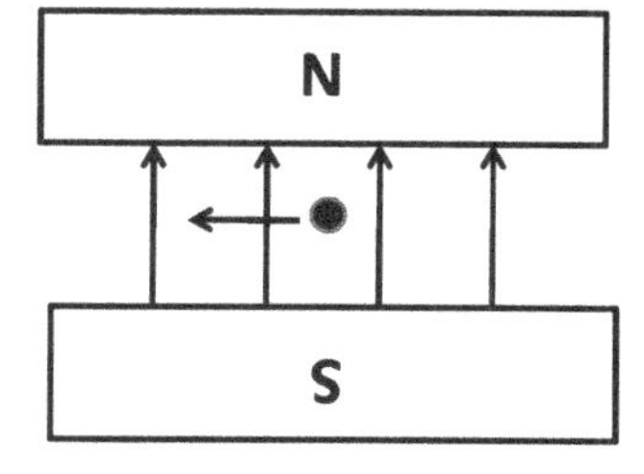

FIGURE 6.7 Dipole magnet.

CHAPTER 7

Vacuum Techniques

Nuclear reaction experiments require vacuum, i.e. pressure of air much less than that of the atmosphere (1,014 mbar). This is because the presence of air molecules can reduce the energy and other properties of charged particles. So normal air will pose problems in doing experiments with ion beams or detecting charged particles. Creating a vacuum is easy, as technology has improved a lot, and one can get vacuum up to 10^{-9} mbar using ultra high-vacuum (UHV) pumps. However, most of the high-vacuum pumps are inefficient in sucking out air at atmospheric pressure. This is achieved by using low-vacuum pumps, which are known as primary pumps. After a primary vacuum is achieved, the high-vacuum pumps are efficient, and a very good vacuum can be achieved and maintained.

7.1 BASIC DEFINITIONS AND CLASSIFICATION

Vacuum is created on earth when pressure of air in a certain enclosed volume is less than the atmospheric pressure of 1,014 mbar. The pressure of a given volume of any gas at a certain temperature signifies the density of its gas molecules at that temperature. Therefore, pressure of a gas can be increased in a volume by increasing the number of its molecules in the given volume. Consider a box of volume 1 litre with a small hole. The box is made of such material that no air can get inside the box except through the hole. In a normal situation, the pressure of air inside this box is 1,014 mbar,

DOI: 10.1201/9781003083863-7

or 760 mm Hg, or 1 atmosphere (1 atm). This is because gases diffuse and maintain equal pressure all around. The hole in the box is closed and a vacuum cleaner is connected to the hole. If the cleaner is switched on, the air molecules are sucked in and pressure inside the box reduces. The pressure cannot equalize because the hole is closed and air molecules from the outside cannot diffuse into the box. If the vacuum cleaner is switched on in the blower mode, a gust of air molecules are thrown inside the chamber and pressure inside the chamber increases to that above atmospheric pressure if it is continued for a long time. If the cleaner is disconnected from the hole, diffusion sets in and the pressure inside the box again becomes 1 atm. In the experimental scenario a sophisticated vacuum chamber (in place of the box), and a vacuum pump (instead of the vacuum cleaner) is used that has a much higher efficiency in creating vacuum than the commercial vacuum cleaner.

Different types of vacuum are commonly classified as

(a) LOW VACUUM (10^{-1} mbar to 10^{-3} mbar)

(b) HIGH VACUUM (10^{-3} mbar to 10^{-7} mbar)

(c) ULTRA HIGH VACUUM (10^{-7} mbar to 10^{-9} mbar)

It is to be noted that the choice of material for a vacuum chamber, its components and the pump to be used depends upon the requirement of vacuum. If the requirement is for a very high vacuum (10^{-9} mbar), then both chamber material and pumps required are very expensive. A non-porous material is suitable for making the chamber. Glass and Perspex are common materials if one needs an optical visibility of the chamber. Stainless steel and aluminium are the most popular choices, as they offer many advantages over other metals or alloys. Stainless steel can be machined easily for a mechanical arrangement and is widely available. Any material that has to be used inside a vacuum needs to be a low vapour pressure material, i.e. substances that will vaporise only when ambient pressure is very low. Non-metals and glues that are often required in the chamber under vacuum should be of low vapour pressure, otherwise they will melt and start emitting vapour when vacuum is created (degassing), thereby spoiling it. Teflon is a good low vapour pressure non-metal. One has to keep in mind that the phase change from solid to liquid and/or gaseous phase occurs at certain temperature and pressure. At lower pressure, the corresponding vaporisation temperature is reduced, and phase change occurs earlier at ambient temperature. Water boils at 100°C, but if kept in a chamber in which a vacuum is created, it starts boiling at ambient temperature (25°C)! This is because the vapour pressure of water gets equal to the small chamber pressure (under vacuum) much more quickly. By heating water in ambient temperature we actually try to increase its vapour pressure and make it equal to atmospheric pressure to start boiling at 100°C. Gaskets also form an important part of vacuum setups where an openable attachment is present in the vacuum chamber. These attachments are useful for accessing the chamber if and when required. Gaskets generally are of two types: metal or non-metal. Metal gaskets are normally made of copper or aluminium, and non-metallic gaskets are made of silicone rubber, elastomer etc. Metallic gaskets are flat whereas non-metallic gaskets are cylindrical in shape. Generally metal gaskets provide better vacuum tightness but get damaged easily each time they are dismantled. As per commercial terminology, the piece of removable component has some standard size, though one can use a custom-made size. The removable part is called a *flange*. The flange fits on the fixed part on the chamber, which has a groove that holds the gasket. The flange is attached with the fixed part using bolts (ISO-F) or clamps (ISO-F) and combination of the two (ISO-KF). There is another type of flange known as conflate or CF type. The size of the flange is described in terms of their bore, such as DN 10, 16, 20, 40, 63, 160, 200 etc. Cleaning of materials to be put under vacuum is best done with ethyl alcohol, carbon tetrachloride etc., as they are volatile and dissolve oils. Elastomer gaskets are best lubricated with vacuum grease that helps these substances to remain soft enough for vacuum tightness.

7.2 VACUUM PUMPS

The most important instrument in a vacuum technique is a vacuum pump. It is like a vacuum cleaner but is more efficient in terms of creating low to high and ultra-high vacuum. There are mainly two types of pumps: (a) primary vacuum pumps and (b) high-vacuum pumps. Another distinction is whether the pump uses oil or operates dry. The low vacuum or primary pumps work in the laminar flow region where density of air or gas molecules is more and intermolecular collisions are more probable than collision of the molecules with the walls of the pump. In the molecular flow region, the high-vacuum pump such as the turbomolecular pump works well. In this region intermolecular collisions are much fewer than the collision of air molecules with the inner walls of the pump. The primary vacuum pumps are therefore efficient at atmospheric pressure and can create a vacuum of about 10^{-3} mbar. The most common type of primary pump are oil rotary pump, roots pump and scroll pump. The latter two primary pumps are dry in the sense that they do not use any oil or liquid to create the vacuum. However, in dry pumps, a small amount of lubricating liquid is used in the bearing. The oils used in the rotary pumps are hydrocarbons, and they release vapour into the vacuum chamber. Dry pumps are therefore used where one wants a very pure environment.

(a) Working principle of an oil rotary pump

The oil rotary vacuum pump (Figures 7.1 and 7.2) is one of the most popular vacuum pump used to create primary or fore vacuum of about 10^{-3} mbar. These pumps are not very expensive. The working principle can be understood from the diagram. The central component of the pump is an eccentric cylindrical rotor (hence the

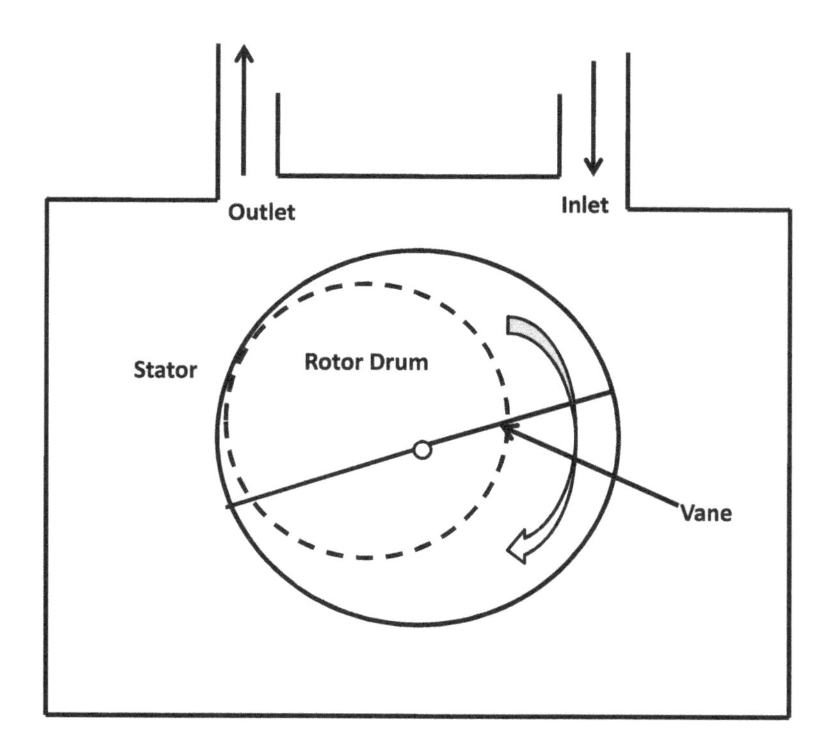

FIGURE 7.1 Oil rotary pump in one stage.

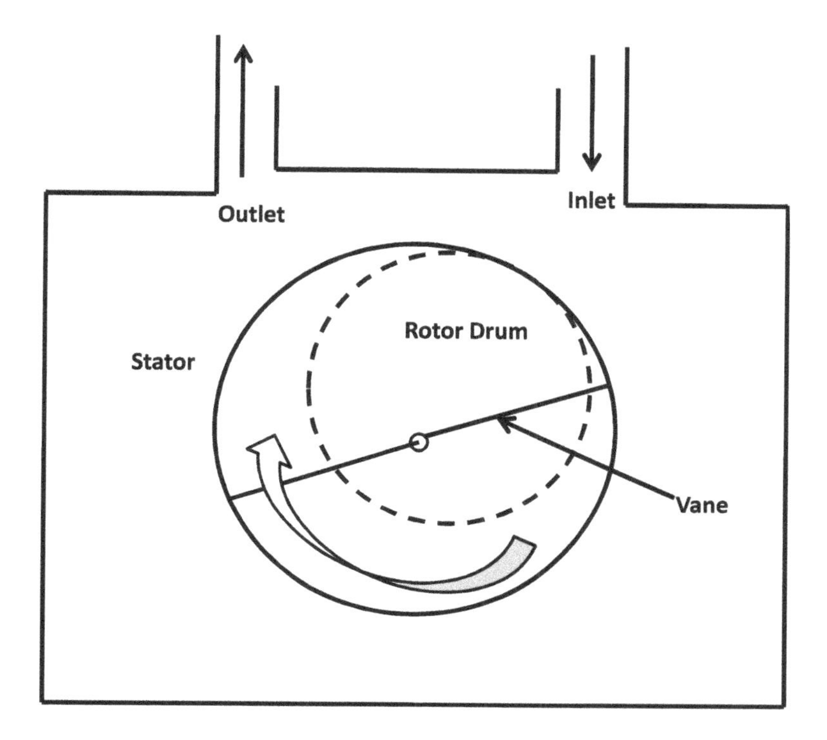

FIGURE 7.2 Oil rotary pump in next stage of compression.

name) that is held by a pair of vanes. These vanes can vary in length due to the spring attached in between. There are openings in the pumps, namely the inlet or the port connected to the chamber to be evacuated and an exhaust (outlet) port to expel the air through a filter. There is oil inside the pump chamber that lubricates the rotor. The rotor is made to rotate by a motor (single or 3 phase) that compresses the air inside the pump chamber and forces it out of the exhaust. The idea is to create a low-pressure area towards the inlet port and high pressure near the exhaust so that air from the vacuum chamber continuously enters the pump chamber and leaves through the exhaust. However, this process is effective as long as pressure of the vacuum chamber is around 10^{-3} mbar. The reason is that at pressure up to 10^{-3} mbar, the flow of air is laminar. This means that layers of air slide with respect to one another. This happens because at these pressures, the density of an air molecule is higher, and so the mean free path is small. As a result, the air molecules collide with each other and tend to move as a bulk. This sort of bulk movement as layers of gas molecules is known as laminar or viscous flow. When pressure falls below 10^{-3} mbar, the density of air molecules becomes less and laminar flow is replaced by molecular flow. In this situation the molecules of air are widely separated and stick to the surface of the vacuum chamber, and as a result become very difficult to sweep out by the method adopted by rotary pumps. The turbomolecular pump works in this situation. This type of pump is like an aeroplane propeller that rotates at very high speed to sweep out the molecules from the vacuum chamber.

(b) Roots pump

It consists of two dumbbell-shaped roots (Figure 7.3) that rotate together to push the incoming air through the outlet. The root pump does not require oil and can be used as a roughing pump.

(c) Scroll pump

This pump consists of two scrolls (Figure 7.4), one fixed and one rotating, that displace air through the middle. The scroll pump, like the roots pump, is a dry pump and can only be used to obtain rough vacuum.

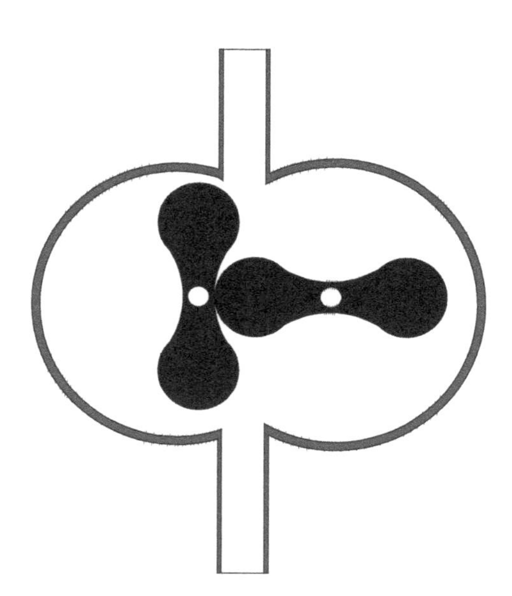

FIGURE 7.3 Roots pump.

FIGURE 7.4 Scroll pump.

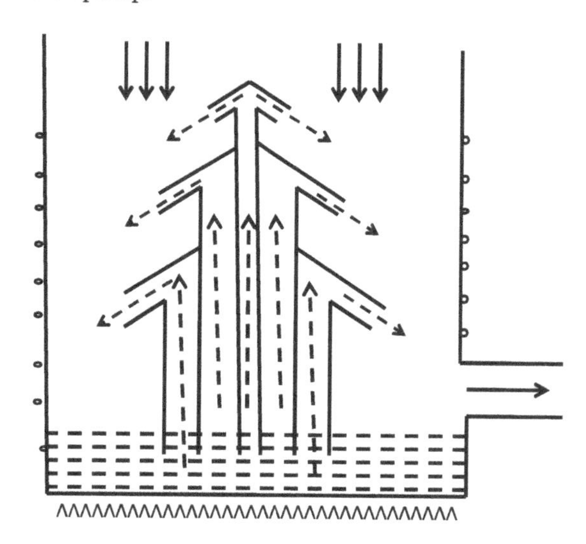

FIGURE 7.5 Diffusion pump.

(d) Diffusion pump

This is an oil high vacuum pump. The oil (hydrocarbon) is heated (Figure 7.5) and vapours are created that move up and come out of the nozzles like a jet. The geometry of the nozzle is such that the vapours flow downward and towards the inner walls of the pump that is cooled by chilled water supply around its outer surface. As the vapour meets an air molecule, the latter is pushed along with the vapour towards the cold wall, where the oil vapour condenses and falls down on the oil reservoir. The number of air molecules is thereby increased in the lower part of the pump, and they are pumped out by a primary pump, as pressure is higher there. At the top (inlet) of the pump the air molecules diffuse to the region near the nozzle (as density of air in this region is continuously decreased by the oil vapour by carrying away air molecules downward) where the oil vapour jet pushes them down towards the outlet. It is to be noted that high vacuum pumps do not operate at higher pressure. The chamber should be evacuated to at least a pressure of 10^{-3} mbar by a primary pump like (a), (b), (c) and then the high-vacuum pumps such as in (d), (e) should be started. As the high-vacuum pump starts creating higher vacuum, the roughing pump acts as a backing pump that throws out as the pressure in the low vacuum region of the high vacuum pump increases.

(e) Working principle of turbomolecular pump

The turbomolecular pump is like a turbine (hence the name) that has an alternate arrangement of stator and rotors as shown in the Figure 7.6. The rotors are fixed with the axle which is connected to the motor, which is again fitted at the low-vacuum side of the pump. The alternating power delivered at 50 or 60 HZ is converted to a higher frequency by a converter so that the pump rotates at a high speed like up to 60,000 rpm (rotations per minute). The molecules hit the rotor blades and are forced down towards the low vacuum side where the rotary pump that created the initial primary vacuum in the chamber drives the air out through its exhaust. The turbo pump has ceramic bearings at the bottom that enable frictionless movement. The lubrication of the bearings is maintained with grease. Modern turbopumps use magnetic bearings that require much less maintenance.

7.3 VACUUM MEASURING/CONTROLLING DEVICES

(a) Vacuum gauges

The vacuum created needs to be monitored by devices known as vacuum gauges. There are separate gauges suited for low and high vacuum measurements and sometimes a combination of both. The Pirani gauge is a typical low-vacuum gauge whereas a Penning or

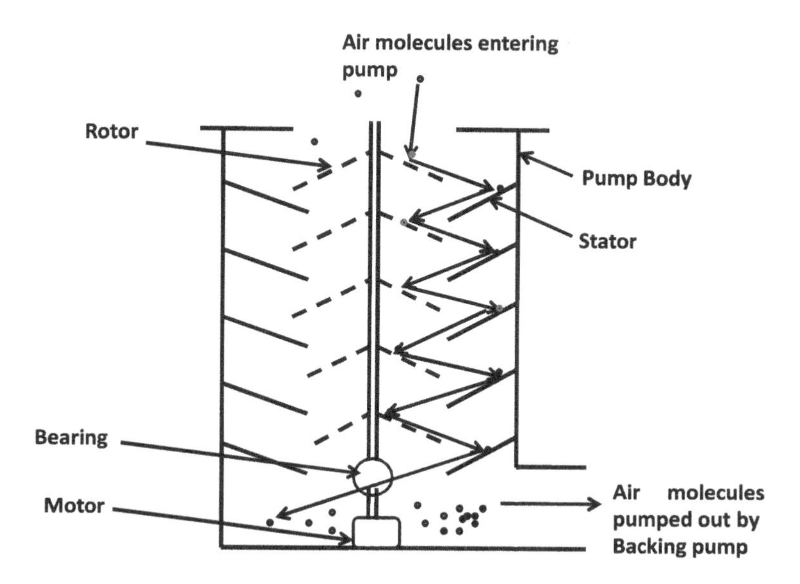

FIGURE 7.6 Turbomolecular pump

Inverted magnetron gauge is suited for high-vacuum measurements. The Pirani works on the principle of thermal conduction of a heater element. The heater element is inserted in the vacuum. The heat of the heater element is carried away by the gas molecules in the chamber to be evacuated. The greater the pressure, the greater is the number of molecules and the amount of heat carried away from the element. The heater element is attached to one arm of a Wheatstone bridge, and when the temperature changes, the current is provided from the power supply to balance the bridge. The increase or decrease of current is proportional to the number of molecules of air present in the chamber. The Penning gauge can measure vacuum ranging from 10^{-3} to 10^{-9} mbar. This gauge is based on ionisation of gas atoms by an electric field and so are also called an ionisation gauge. Therefore, the larger the number of gas atoms (higher the pressure of air), the higher is the ionisation, but at very low pressure the number of molecules are very few, and so the ionisation is not sufficient for detection. In order to create more ionisation of the gas atoms, the electrons created by the electrical discharge are made to travel a longer path (so that the electrons gets a much higher chance to meet the few atoms remaining at such low pressure and ionise them). In this way the ionisation current is scaled up and very low pressure can be measured.

(b) Valves

Valves are an integral part of a vacuum system. Valves can be used for separating two high-vacuum regions, the high-vacuum and low-vacuum regions, and air and a low-vacuum region. There are valves that are used for letting the air inside a vacuum very slowly (vent or leak valves). As air has moisture, dry nitrogen or argon is used for venting so that good vacuum can again be retained in a short time. Valves can be operated pneumatically (using high pressure air or gas) or manually.

A basic vacuum generation setup is shown in Figure 7.7. The chamber to be evacuated has three openable ports. One port is connected to a

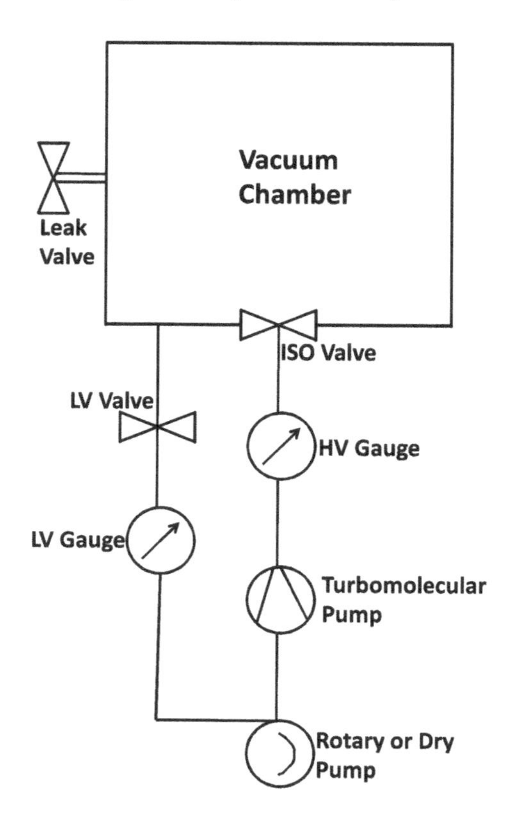

FIGURE 7.7 A basic vacuum generation setup.

low-vacuum valve, a low-vacuum gauge and a pump. This branch is sometimes known as a roughing line, as it is used for creating low vacuum inside the chamber. Once the low vacuum condition is achieved, the low-vacuum valve is closed and the high-vacuum isolation or gate valve is opened and the turbomolecular pump is powered. The low-vacuum pump now acts as a backing pump for the turbomolecular pump. Once high vacuum conditions are achieved, the chamber can be brought to atmospheric pressure by introducing air using the leak valve.

CHAPTER 8

Radiation Detectors

Radiation in the form of charged nuclei, neutrons and gamma rays is detected by nuclear radiation detectors. They are microscopic particles and can also be thought of as waves that have wavelength beyond the visible spectrum. Thus they are released by some interaction they produce in the detector medium. The working principle of these detectors can be broadly classified into those that depend on (a) ionization produced by the radiation in the detector medium and (b) scintillation or light (not necessarily in the visible range) produced in the detector medium. The detector media are mostly either a solid or a pure gas, though liquid scintillators are used in some very rare situation.

8.1 CLASSIFICATION AND TYPES OF RADIATION DETECTORS

The classification of detectors depends upon what particle is detected and therefore how the particle is likely to interact with the detector medium chosen. Charged particles interact with the matter (in the detector medium) by ionization. A radiation has two important parameters in terms of its detection, namely its dynamic property, i.e. kinetic energy, and static parameters such as mass, charge, spin etc. The kinetic energy is a dynamic parameter, as it changes as the radiation passes through the detector. On the other hand, the static properties do not change. A very

DOI: 10.1201/9781003083863-8

important quantity in nuclear physics measurement is the particle's energy. So a detector should ensure that the particle loses energy in the detector and somehow measure this energy loss. The most convenient way is to detect if some currents or charges are generated by the interaction, as we have a lot of available electronic instruments to detect charges and currents. The process of *ionization* does this directly, whereas in the process of *scintillation* it has to be done indirectly, as will become evident later.

Ionization is the process by which an atom loses electrons by collision with other atoms that make the detector medium. In this process, the detector atoms also lose electrons, and in the whole process the incident particle loses its energy, as the electrons are bound in the atoms and need some energy to dispose of them from their shells. Charged nuclei easily lose their energy by ionization as they pass through a medium, and this energy loss can be accounted by the Bethe–Bloch theory. The details of the theory are complicated, but the specific energy loss (energy loss per unit length of travel) of an atom with charge state z (the number of electrons that it has lost from its neutral condition), mass m and energy E is given as

$$\frac{-dE}{dx} \propto \frac{mz^2}{E} \tag{8.1}$$

where dx is the element of length the particle has travelled through the detector. Integrating over the entire length (say l) of the detector, the total energy lost is

$$E = \int_0^{*l} \frac{-dE}{dx} dx \tag{8.2}$$

So it is necessary that the detector should be long enough to stop the particle. The distance over which the particle looses all its kinetic energy is known as its *range (R)*. So for energy measurement of the particle, the condition $l \geq R$ should be satisfied. The particle range can be calculated by using programs like SRIM with input for the particle mass, charge state, energy and the detector medium. The charge state is an important quantity, as the energy loss depends on the second power of it. A nucleus is an atom that has lost all its electrons. In case of a proton, a hydrogen atom can lose a single electron, and so it has a +1 charge state only. However an oxygen nucleus is an atom with +8 charge state but can also be in other charge states such as +6, +7, +1 etc. and is capable of losing energy in different ways depending upon the charge state. The gamma rays lose their energy completely by photoelectric effect or photoionization process only. However, a gamma ray can undergo other processes like Compton scattering and/or by production of electron–positron pair where its energy is partially lost in the detector medium. As photoelectric cross-section increases with the atomic number of the detector medium, high Z materials are preferable for gamma detectors. Measurement of neutron energy is not possible by ionization, as the neutron does not lose any energy by interaction with the electron and instead interacts with the nucleus of the atom. Nuclear scattering with lighter atoms like hydrogen, ^{3}He, lithium or boron will change the energy of the neutron very quickly with the angle of scattering. A better way is to use the time-of-flight method described in the chapter, as the time of occurrence of a neutron can be well determined by some of the neutron detectors.

8.2 DETECTOR EFFICIENCY AND RESOLUTION

The most important quantities that are measured in nuclear reaction experiments are the cross-sections and the kinetic energies and time of occurrence of a charged nuclei or a neutron. In the measurement of the cross-sections one has to record the number of particles of interest with the detector (the yield in the chapter). These particles come out with a certain angular distribution (in fact, they are spread over a solid angle), but often the detector's active areas are much smaller and cover a small percentage of the actual 4π solid angle. Thus the geometric efficiencies are smaller for smaller-area detectors and larger for larger-area detectors. Recently, large-area detection geometry is being created by installing multi-detector arrays. Large-area single detectors instead of multiple detectors can be used, which can increase the geometric efficiency, but as nuclear reaction cross-sections are angle dependent, care must be taken to increase the granularity of these detectors so that a contribution from several angles is not counted as a contribution from a single angle. However, achieving such high granularity using a single large-area detector is very difficult technologically. Besides the geometric efficiency, the intrinsic efficiency of the detector also must be considered. This is the ability of the detector to produce an electrical signal as a particle interacts with the detector medium. When a detector is not being able to create a signal, such failures can occur if a radiation can interact with the detector medium through multiple processes, but only one of them helps in detection of interest. In such a case the detector is not 100% efficient. This is best understood by comparing the measurement of energy of a charged particle and gamma ray by semiconductor detector. If 100 charged particles of different energy fall on the detector, all of them interact to give an electrical signal, and intrinsic efficiency of a charged particle detector is 100%. The gamma rays, on the other hand, can deposit energies in three different ways, namely photoelectric effect, pair production and Compton scattering, and probability of each process depends on energy. In order to measure the energy of the gamma ray, full energy deposition is required, so it is very unlikely that all gamma rays that fall on the detector will deposit energy by photoelectric effect. So efficiency of gamma detectors is always less than 100% and depends on energy. The efficiency (ε) of a detector is therefore defined as

$$\varepsilon = \varepsilon_g \times \varepsilon_i \tag{8.3}$$

where g and i indicate the geometric and intrinsic efficiencies, respectively, and are considered independent of one another. Efficiency is normally expressed as a percentage.

Resolution is a parameter that signifies the quality of a detector in terms of measuring any physical quantity precisely, such as energy, angle and time of occurrence of an event in nuclear reaction. For example, in some experiments (nuclear spectroscopy) one has to measure the energy states of a nucleus that may be very close to each other. The measurement of this state by any detector does not produce a delta function, as the measurement has some uncertainty. This uncertainty arises mainly from the statistical fluctuations in the number of electrons produced by the interaction and the electrical noise of the associated electronic instruments that process the signal. The smaller the uncertainty, the better the resolution of the detector. Since the statistical process can be energy dependent, the resolution is normally expressed in % for a particular energy. The timing resolution depends upon how quickly the detecting process creates the signal from the interaction. This of course depends upon the type of detector and the incident radiation.

8.3 GAS DETECTORS

Gas detectors use gas as the detecting medium, hence the name. When a radiation falls on the gaseous medium, it creates electron ion pairs by ionization of the gas atoms. This is in contrast to the electron hole pairs created in a solid medium, as discussed in the next section. The number of electron ion pairs (n) is given by the relation,

$$n = \frac{E}{W} \tag{8.4}$$

where E is the energy of the incident radiation and W is the threshold energy for ionization. Using a typical value of an α particle emitted in a radioactive decay of 6 MeV and threshold for gas of about 30 eV, the number of electron ion pair is 200,000. In solids the threshold energy is an order less and the number of electron hole pairs is an order more. As the statistical error $\frac{1}{\sqrt{n}}$ in n will be much lower for solid-state detectors, the energy of the radiation which is directly proportional to the number of electron-ion or electron-hole pair produced, the energy resolution in solid-state detectors will be better than in gas detectors. However, gaseous detectors have several advantages over their solid-state counterparts. The foremost is that they are inexpensive and not prone to radiation damage. The thickness of gas detectors can be varied as one chooses by varying the gas pressure. The basic principle of operation is that a pure gas is enclosed in a volume where, if radiation falls, its energy is lost by ionization of the detector gas atoms. Electrons and ions are therefore formed along the track of the incident radiation. A very basic design of a gas detector is shown in Figure 8.1. The two electrodes are connected to a regulated DC power source. The electronic signal is collected at the anode through a capacitor to block the DC voltage (for DC frequency is zero and reactance of a capacitor becomes infinite) interrupting the signal (transient as it is varying with time).

The basic features of the detector are explained through a plot of the number of electrons collected at the anode as a function of the applied DC voltage (Figure 8.2) across the anode and cathode. When there is no applied voltage (point A in Figure 8.2), no or very few electrons are collected, as most of the electron-ion pairs formed recombine with each other to form neutral atoms. As the DC voltage is increased, the number of electrons collected becomes more compared to the loss due to recombination. Thus the number of collected electrons increases linearly (region AB in Figure 8.2) with increasing voltage. But after a certain bias voltage, all the electrons that are produced by primary ionization are collected and recombination is overcome. After this voltage the current does not change

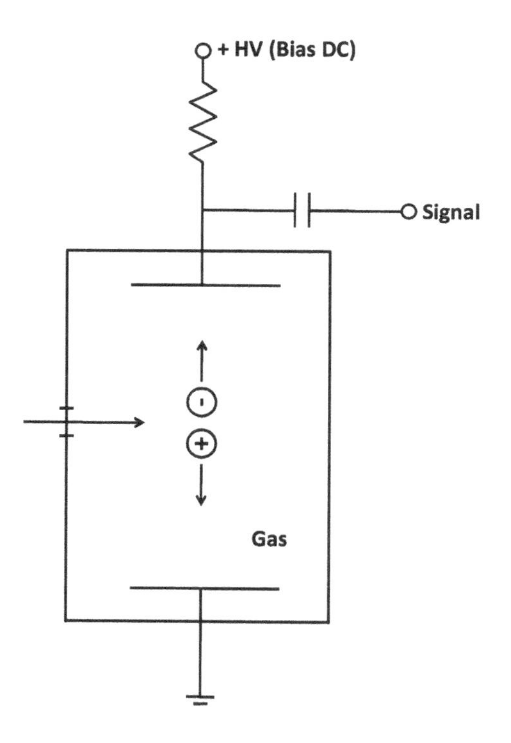

FIGURE 8.1 Basic design of a gas detector.

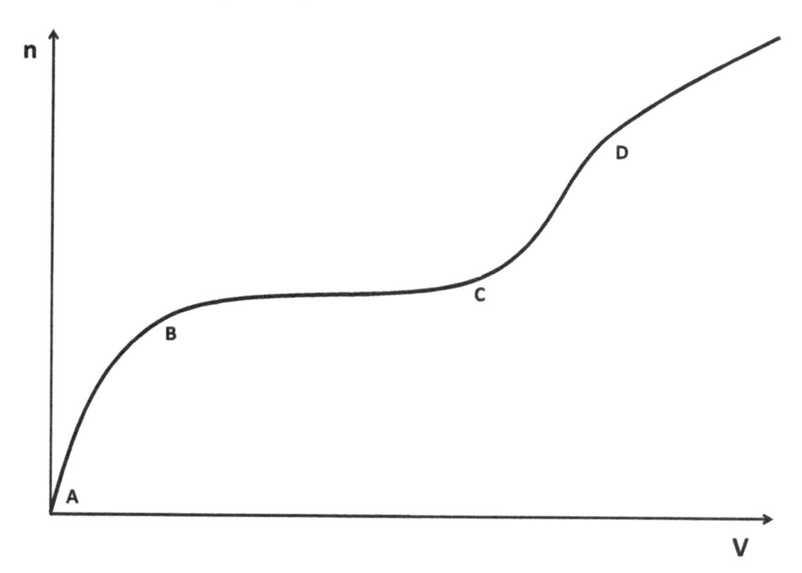

FIGURE 8.2 Different operating regions of a gas detector.

because the number of electrons does not increase (saturation region BC in Figure 8.2). The number of electrons again increases after a certain bias voltage, and this is due to secondary ionization initiated by the electrons created from primary ionization rather than from the incident radiation. Secondary ionization is a process in which an electron produced from primary ionization ionizes a gas atom in the detector. Probability of secondary ionization increases only when the electrons from primary ionization have gained enough energy from the increased bias voltage. The number of secondary electrons produced is proportional to the number of primary electrons, and the region CD shown in the Figure 8.2 is known as the proportional region. If bias voltage is further increased, the number of secondary electrons becomes independent of the number of primary electrons and an avalanche is created (beyond point D).

Different gas detectors operate in different regions of the figure. The ionization chamber works in the saturation region, the proportional counter in the proportional region, and the Geiger Muller counter in the avalanche region. As both in saturation and proportional region the number of electrons produced is proportional to the number of primary electrons, they are useful for measuring the energy of incident radiation, as number of primary electrons is directly proportional to the energy of the radiation (equation). In the Geiger region it is only possible to detect the radiation, but the electric field and the gas pressure play an important part in the various modes of operation of gas detectors.

Some of the most common gas detectors are:

(a) Ionization chamber (IC): These detectors operate in the saturation region, i.e. they rely only on primary ionization. As such, the electric field is normally generated by parallel plates either in axial or transverse mode (Figure 8.3). Charged particles, especially heavy ions that lose energy profusely in gas, are most suitably detected by ICs. The electric field for a parallel plate is $E = \frac{V}{d}$ where d is the separation between the parallel anode and cathode plates. In the ionization chamber there is no secondary ionization, and the ratio of electric field (E) to the gas pressure (P), E/P is kept small (1–2 V/cm/Torr). This type of detector works on the amount of electrons produced by primary ionization, so heavy ion detection is most suitable with this

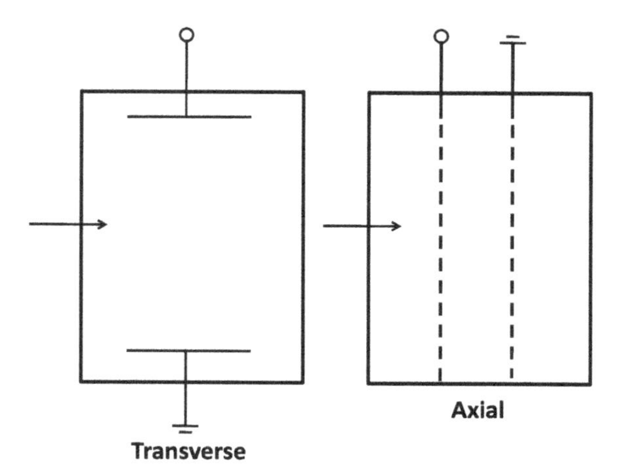

FIGURE 8.3 Transverse and axial field geometry.

detector, as the heavy ions produce a larger energy deposition than lighter ions, owing to their larger charge states.

(b) Single-wire gas detector: A proportional counter is a gas detector that works on the basis of secondary ionization process. In these detectors, a higher electric field is required, which is achieved by using a thin wire and a cylindrical detector. The thin wire runs through the axis of the detector and acts as the anode.

The detector body is grounded and acts as the cathode (Figure 8.4). In this cylindrical geometry, the radial electric field at any point r from the centre (on the detector axis) is given by

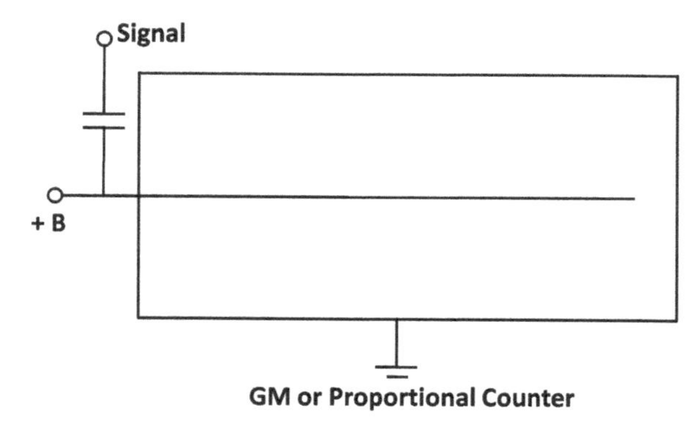

FIGURE 8.4 Single-wire gas detector.

$$E = \frac{V}{r} In \frac{b}{a} \tag{8.5}$$

where b is the radius of the hollow cylindrical detector body and a is the radius of the thin wire in the middle. The electric field increases as the electrons created by primary ionization come closer to the wire. So the maximum secondary ionization occurs very close to the wire and is collected as a strong signal. A thin wire (very small a) enables the electrons to come very near the axis, and thus larger fields can be achieved. However, if the bias voltage on the anode is not made very high, the number of secondary electrons produced is proportional to the number of primary electrons. As such, these detectors can be used for particle energy detection. The type of gas used and its pressure are also of importance for operation in the proportional region. The gas P10 (10% methane and 90% Argon) is a very popular proportional counter gas. If very high voltage is used in a single-wire gas detector, then the avalanche region is reached where the secondary ions give rise to series of ionizations and the number of electrons produced has no correlation with the number of primary electrons. This is the operation region of the Geiger Mueller (GM) counter, which is very good at detecting the presence of radiation but is unable to measure its energy.

(c) Multi-wire detector: In recent times, instead of a single wire, multiple thin wires are used in gas detectors to obtain better efficiency. This is because a larger solid angle can be covered by such detectors, and the anode and cathode are made in the form of wire planes (Figure 8.5).

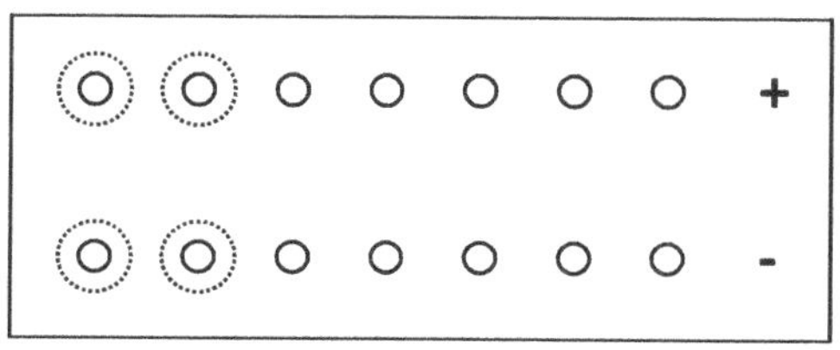

FIGURE 8.5 Multi-wire gas detector.

The E/P ratio in such a detector is kept very high compared to a single-wire detector and is of the order of 200–300 V/cm/Torr. Such high values of electric field to gas pressure ratio is brought about by maintaining a low pressure of gas (2–5 Torr) in the chamber. Such high value of E/P introduces a gas gain in the electron multiplication besides the multiplication in the vicinity of each wire as in the single-wire detector. Owing to its larger area of coverage by multiple wires, these detectors are good for position detection, and since the amplification occurs very close to the wires, the timing property of the detector is also very good.

8.4 SOLID-STATE DETECTORS

Both semiconductor and scintillator detectors are important to discuss in this category. Semiconductors like silicon and germanium are very popular as radiation detectors because they are very compact and offer excellent energy resolution for charged particles and gamma rays. Both silicon (Z = 14) and germanium (Z = 34) have four electrons in the outermost orbit and form covalent bonds with four such atoms sharing the four electrons in a crystal lattice. In a solid, there is a high density of electrons, and so if they are free from the atomic binding, they can form a large number of levels that form a continuous band-like structure known as the conduction band. The electrons that are bound in the form of bonds reside in definite energy levels obeying Fermi statistics. The levels near the Fermi energy come closer and are known as the valence band. The energy required to push bound electrons from the valence band to the conduction band is known as the band gap, and for silicon and germanium it is of the order of eV. This is small enough to excite electrons by the thermal energy available in room temperature situation. However, the number of thermally excited electrons in a semiconductor depends exponentially on the bandgap E_B and the temperature T^0 in Kelvin. Thus, though the bandgap of silicon is slightly greater than that of germanium, the number of electrons thermally excited is much higher. The pure semiconductors such as silicon and germanium conduct a very small current and are not suitable as radiation detectors. This is because if radiation undergoes interaction with the bond-making electrons, the lattice structure will be damaged and electrons may be pushed to the conduction band haphazardly without any relation to the energy of the incident radiation. This is because as electrons are removed from their bound shells, they create vacancies called holes. In absence of any electrical field, the electrons produced have a large probability of recombination with the holes. Thus, the electrical signal will be very weak and will have no correlation with the energy of the radiation, and the purpose of the detector will not be served at all. The solution to this situation is to dope the pure (intrinsic semiconductor) with n- (or p-) type impurity. An n-type impurity is a pentavalent atom, and a p-type impurity is a trivalent atom. If a pentavalent atom replaces a germanium or silicon atom and forms the octet by forming covalent bonds with four other germanium or silicon atoms, one electron becomes very loosely bound and can easily be excited to the conduction band. If a trivalent atom is used, then an electron is deficient or a hole is created. Thus, in an n-type semiconductor, electrons are in excess, and in a p-type semiconductor, holes are in excess. Again, n-type and p-type semiconductors can have higher conductivity than intrinsic semiconductors do, but they are not useful as detectors because without an electric field, the electrons elevated to the conduction band can come back and recombine with the holes, producing a very weak signal.

A radiation detector can be built from a combination of an n-type and p-type material fused together. This is well known as a p-n junction diode, which has rectifying properties in the sense that it conducts a current only when the diode is forward biased (i.e. the p-type side is at a higher electrical potential than the n-type side). It conducts a very small current in the reverse bias mode till breakdown voltage is reached. As the p and n materials are fused, there is a net imbalance of free carriers on both sides. Electron concentrations being more of the n-type try to diffuse towards the p-side, and holes try to diffuse towards the n-side. As they move towards each other, they begin recombining, and a charge-free region is created. If the doping concentration is low, the recombination takes place over a greater distance from the junction, and vice versa. A carrier-free region is created over a distance on either side of the junction due to the recombination, which is known as a depletion region. If doping concentration of the p-side (say p^+) is significantly increased, the depletion region will be much more on the n-side than on the p-side. As electrons leave the n-side, they leave holes, and holes leave electrons on the p-side. As a result, an electric field builds up across the carrier-free depletion region, which prevents further diffusion. The depletion length can be a few micron to a few millimetres. A reverse bias will increase the depletion depth by slowing the electrons and holes so that they travel greater distances for recombination. The depletion region acts as a region of high resistivity, like a capacitor with the reverse bias potentials, to attract the electron-hole pairs across the bulk material. The charging up of the depletion region due to the potential is like the charging up of the gas volume in the gas detector. The capacitance of the depletion region will change with the reverse bias voltage and is minimal at the highest reverse bias voltage. The time taken by all the electrons collected for a particular interaction is called the rise time of the signal (of the order of nanosecond) and is dependent on the property of the semiconductor material. Each time an incident radiation interacts with the material inside the depletion region, electron-hole pairs are formed. The electric field due to the space charge (ions) and the applied reverse bias pulls these electrons towards the electrodes before they start recombining with the holes. A current pulse is generated by the flow of charge, which determines the total charge flow. The current pulse is an integration of the several charge pulses, each of whose amplitude is proportional to the energy deposited by incident radiation. The information on each energy level deposited is required in spectroscopy measurement, and this is done by pulse mode detection. The electronic processing of the charge is discussed in the next section.

Three types of diode detectors are commonly used in nuclear reaction experiments: (1) surface barrier, (2) ion implanted and (3) lithium drifted. When two metals of different work functions are kept in contact, electrons in the metal with a smaller work function are at a higher energy level (smaller work function metals have electrons with lesser binding energy, and vice versa) and transfer electrons to the other metal (say B), thereby creating an electrostatic potential or electric field. This continues until the Fermi levels align themselves. As a result, no more electrons flow. A similar effect is observed at a metal–semiconductor junction, as the band structure is modified due to the difference in Fermi levels of the two materials at first, which comes to equilibrium until a contact potential is developed. If the work function of the metal (Figure 8.6) is higher (gold–silicon), then a Schottky junction is formed that acts like a p-n junction diode and can function as a surface barrier detector. This is because the barrier (V_B in Figure 8.7) increases under reverse bias condition (considering the silicon in the figures as an n-type material with positive bias on it) and reduces the flow of electrons, whereas in the forward bias the barrier is reduced and electrons flow more with

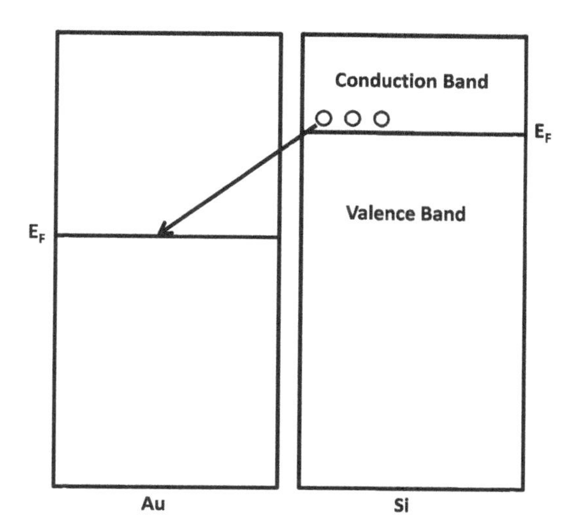

FIGURE 8.6 Metal (Au)-semiconductor (Si) before contact.

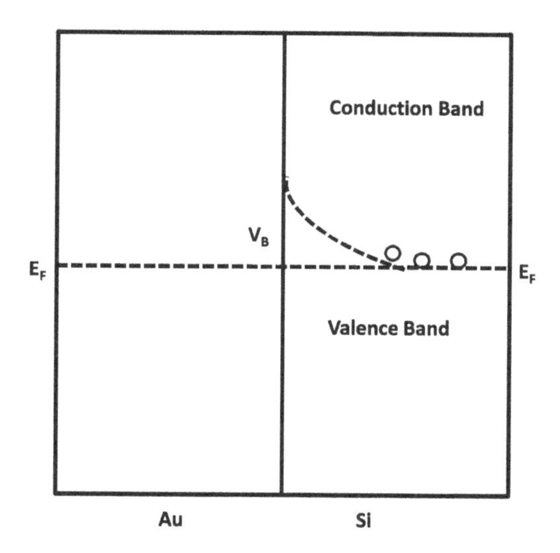

FIGURE 8.7 Metal (Au)-semiconductor (Si) Schottky contact.

increasing bias. In order to have an electrode to bias the surface barrier detector, another metal–semiconductor junction (Figures 8.8 and 8.9) is formed where the work function of the metal is smaller. This is known as an Ohmic contact, and aluminium–silicon forms such a contact. Ion-implanted detectors are manufactured by implanting either boron or phosphorous beams in silicon. Lithium is a trivalent element and is used

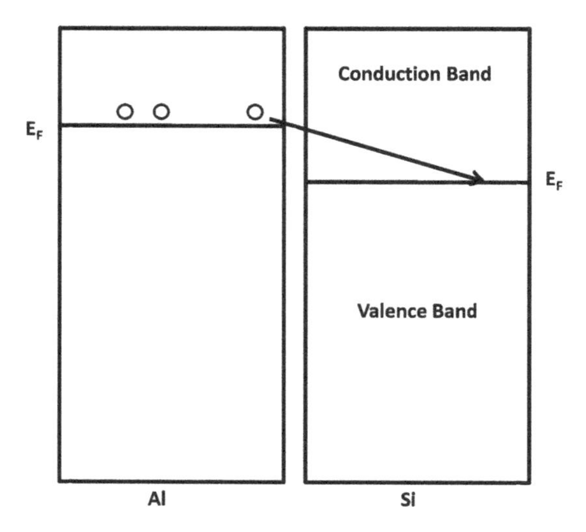

FIGURE 8.8 Metal (Al)-semiconductor (Si) before contact.

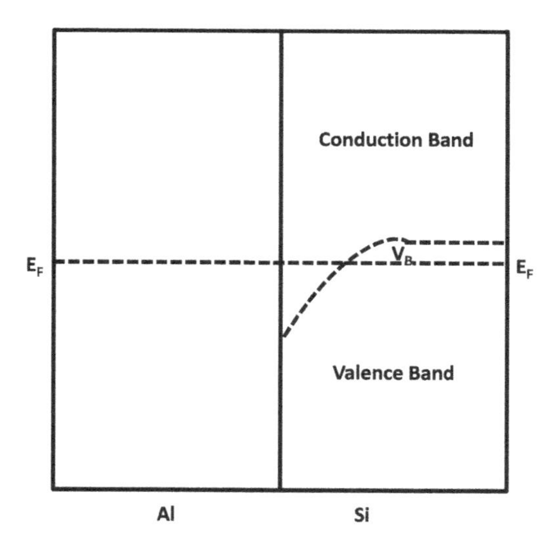

FIGURE 8.9 Metal (Al)-semiconductor (Si) Ohmic contact.

to form a very thick depletion layer of up to 5 millimetres. These detectors are known as lithium drifted detectors.

Scintillation detectors detect the nuclear particles indirectly. These are materials that give out a photon when radiation loses energy in them. Instead of an ionization, if an excitation occurs, an electron bound in the atoms of the scintillator material is excited to a higher energy level, which when exiting to the lower level, gives out an electromagnetic wave. These photons then fall on a photocathode that emit electrons which convert the light signal to an electrical signal. The latter device is known as a photomultiplier and has a window that interfaces with the scintillator. This window should have high transmission for the photons. This depends upon the wavelength of the scintillation photons. In case of visible light, a glass window can be used, but for UV photons, transmission in glass is reduced, and quartz or fluorides like MgF2 and LiF are good window materials. Sometimes wavelength shifters are used so that wavelengths of photons are shifted to the visible region and a glass window can suffice. Wavelength shifters can be either put on the window or as an impurity in the scintillator material. For example, NaI (sodium Iodide) is an inorganic scintillator that is very popular for radiation detection, particularly for gamma detection. Thallium is doped in sodium iodide to shift the wavelength of the original scintillation light to the visible region. The photocathode should have high light conversion efficiency, and the electron output should be proportional to the light falling on the cathode.

CHAPTER 9

Electronics and Data Acquisition System

The basic output from a radiation detector used in a nuclear reaction experiment is an electric charge, varying over a period of time. The charge can be converted to a voltage or a current pulse. A pulse is a voltage or a current that is varying over time. This time variation comes from the charges deposited by a radiation in the detector that gradually charges the detector capacitance. If one connects a measuring device (such as an oscilloscope) to study the pulse, nothing is seen because the pulse needs some processing, for which the knowledge of the electronic setup is required.

9.1 ANALOG PULSE FORMATION FROM RADIATION DETECTORS

The basic electronic setup starts with the observation of the raw pulse (signal) from the detector. In order for the signal to be observable, impedance of the detector and that of the oscilloscope should be matched, otherwise the signal loses strength. The detector impedance is typically a few MΩ, and so the oscilloscope input impedance is of the same order. Despite that, the signal is very weak and hardly noticeable over the noise level (typically few tens of millivolt). A small amplification is therefore required, as well as impedance matching, which is provided by a preamplifier. The preamplifier also takes care of other necessary devices to be attached after it. There

DOI: 10.1201/9781003083863-9

are different types of preamplifiers that are used depending upon the requirement.

The time-varying charge can carry three important types of information, namely the peak or amplitude of the pulse; its rise time; and the area under the charge versus the time curve (which is the total current). Each of these pulses (an existence of voltage or current over a short interval of time) corresponds to a particular event that has fallen on the detector. In practice, there may be a huge number of events, and in most cases a measurement will require processing of these individual pulses for various experiments.

The detector (gaseous and semiconductor), as has been discussed earlier, acts as a capacitor (C_d) with two collecting electrodes to which an electric field is applied. A radiation pulse, after interaction with the detector medium, creates electron-ion (for gases) and electron-hole (for semiconductors) pairs. Each charge induces an equal and opposite charge on the collecting electrodes depending upon the charge's position with respect to the electrode. As the charge moves towards the opposite electrode (giving rise to the current), the induced charge and distribution change. The amount of induced charge can be easily calculated using Gauss's theorem. The detector capacitor charges up with the induced charge and goes on charging until the entire charge created is collected. The time taken for all the charges to reach the electrodes is the charge collection time (τ_c). If the detector has an effective resistance R_d, the nature of the pulse will depend upon the time constant R_dC_d. The detector generates a current pulse that exists during the charge collection time and can be represented as $i(t) = i_0 e^{-\frac{t}{\tau_c}}$. This detector current pulse flows through the detector capacitance and a resistor in parallel. The current across the resistor is $\frac{v(t)}{R_d}$ and across the capacitance is $C_d \frac{dv(t)}{dt}$, so that

$$i(t) = \frac{v(t)}{R} + C_d \frac{dv(t)}{dt} = i_0 e^{-\frac{t}{\tau_c}} \tag{9.1}$$

Solving for $v(t)$ in the linear differential equation and boundary condition $v(0) = 0$, one gets

$$v(t)=\frac{i_0}{C_d\left(\frac{1}{RC_d}-\frac{1}{\tau_c}\right)}\left(e^{-\frac{t}{\tau_c}}-e^{-\frac{t}{RC_d}}\right) \tag{9.2}$$

If the time constant is very small ($\frac{1}{RC_d}$ is large), at very small times ($t \ll \tau_c$) $\frac{1}{RC_d}t=0$ with $\frac{1}{RC_d}-\frac{1}{\tau_c}=-\frac{1}{\tau_c}$ and

$$v(t)=i_0R_d\left(1-e^{-\frac{t}{RC_d}}\right) \tag{9.3}$$

$$=\frac{i_0\tau_cC_dR}{\tau_cC_d}\left(1-e^{-\frac{t}{RC_d}}\right) \tag{9.4}$$

$$=\frac{Q}{C_d}\frac{C_dR}{\tau_c}\left(1-e^{-\frac{t}{RC_d}}\right) \tag{9.5}$$

At larger time,

$$v(t)=\frac{Q}{C_d}\frac{C_dR}{\tau_c}e^{-\frac{t}{\tau_c}} \tag{9.6}$$

So the voltage pulse exhibits almost the same behaviour as the current pulse except at very small energies. But the amplitude of the voltage pulse is less by a factor of $\frac{RC_d}{\tau_c}$ of the maximum expected amplitude, i.e. $\frac{Q}{C_d}$. Thus, a small time constant is not suitable for energy measurement, as the total charge is not obtained in this mode. If a large time constant is used, then at small times ($t \ll RC_d$),

$$v(t)=\frac{Q}{C_d}\left(1-e^{-\frac{t}{\tau_c}}\right) \tag{9.7}$$

and at larger times,

$$v(t) = \frac{Q}{C_d} e^{-\frac{t}{RC_d}} \tag{9.8}$$

At a large time constant, the maximum voltage can be achieved with the rise time determined by the charge collection time and the decay time by RC_d.

Usually the pulse amplitude from a detector is quite small (of the order of few tens of mV), and amplification is required. Moreover in semiconductor-type detectors, the capacitance of the detector (C_d) can change with the applied reverse bias and can change the amplitude that gives a measure of the energy. This undesirable effect can be negated by using a charge-sensitive preamplifier. A preamplifier is an OP AMP (Operational Amplifier) with a very large input impedance and small output impedance, and the gain (A) is very large. A feedback capacitor (C_f) is used at the inverting terminal. As no charge flows through the OP AMP (but only flows through the feedback path),

$$C_i(V_+ - V_-) = C_f(V_O - V_-) \tag{9.9}$$

$$C_i(0 - V_i) = C_f(V_O - V_i) \tag{9.10}$$

$$C_i = C_f(A+1) \tag{9.11}$$

by substituting $V_O = AV_i$. The input capacitance C_i appears in parallel to the detector capacitance C_d. Thus,

$$v_{in} = \frac{Q}{C_d + C_i} \tag{9.12}$$

$$= \frac{Q}{C_d + C_f(A+1)} \tag{9.13}$$

and the output voltage is

$$v_o = Av_{in} \tag{9.14}$$

$$= A\frac{Q}{C_d + C_f(A+1)} \tag{9.15}$$

$$= A\frac{Q}{C_f(A+1)} \tag{9.16}$$

$$= \frac{Q}{C_f} \tag{9.17}$$

The output signal is thus dependent on the fixed feedback capacitance C_f, and the variation in detector capacitance does not affect the output. The above relation holds, as the gain of the OP AMP is very large.

However, the amplitude of the output signal of the preamplifier is still small and has a long tail owing to the large time constant. If another pulse appears before the pulse decay has been completed, the pulses ride on one another, leading to distortion of the pulse amplitude and hence failing to measure the energy correctly. This effect is known as a pile-up. In order to amplify the signal further and resolve the pile-up, a shaping of the preamplifier pulse is performed. This is done by a spectroscopy amplifier used for energy (spectrum) measurement. The amplifier, in contrast to the preamplifier, has a small input impedance, thereby allowing impedance matching with the preamplifier (maximum power transferred when impedance is matched). The amplifier mostly provides a Gaussian shaping of the preamplifier pulse. This Gaussian shaping is done by a RC-CR^n filtering circuit. This means that the signal from the preamplifier is differentiated once and integrated n times. If n is kept greater than or equal to 6, it is enough to achieve the Gaussian shape. This shaping also helps to improve the signal-to-noise ratio.

9.2 ANALOG TO DIGITAL CONVERSION

The amplifier signal needs to be digitised to be read into a computer for convenient analysis. The signal from an amplifier is analog in nature and can be digitised by a device known as an analog-to-digital converter or ADC. The sensitivity of the ADC depends upon the number of bits it can handle. A 2K ADC will divide an analog signal into 11 bits or 2^{11} (2,048, not exactly 2,000) channels. So, the higher the bits, the better the resolution. Usually ADCs accept a low voltage within 10 V or sometimes within 2 V. ADCs are of three popular types, namely (a) counter, (b) successive approximation and (c) flash. The first two types are sequential (or series) in nature whereas the flash type is simultaneous (or parallel) in nature. These aspects can be understood in terms of some examples. Let us take an analog signal of amplitude 4 V which is to be digitised by a 10-bit ADC. An N-bit ADC has N registers for defining a parameter through binary numbers 0 and 1. So a 2-bit ADC will have two registers, 8-bit ADC eight registers, 10-bit ADC 10 registers, and so on. A 2-bit ADC can have 2^2 (4) possible combinations of these numbers, namely 00, 10, 01 and 11. Similarly, a 10-bit ADC can have 2^{10} combinations, and in general an N-bit ADC can have 2^N combinations. The first bit or the register on the extreme left is called the most significant bit (MSB), and the one at the extreme right is known as the least significant bit (LSB). The reference voltage usually is 10 V. The input voltage of 4 V is compared with $V_d = \frac{V_{ref}}{2}$. As 4 V is less than 5 V, the MSB is assigned 0. Next V_d is set to $\frac{V_{ref}}{4}$, i.e. 2.5 V. Now the input voltage is greater than V_d, so the second bit is set to 1. The next V_d is $\frac{V_{ref}}{4} + \frac{V_{ref}}{8}$, i.e. 3.75 V. As V_{in} is less than 3.75 V, the third bit is set to 1. (Point to note is that when the bit is 1, the next V_d is set considering the previous V_ds for which bit value was 1, whereas if the bit is 0, the next V_d does not consider the earlier V_d). The V_d for the fourth bit is therefore $\frac{V_{ref}}{4} + \frac{V_{ref}}{8} + \frac{V_{ref}}{16}$, i.e. 4.375. Again V_{in} is less than V_d, and so the fourth bit is 0. The V_d for the fifth bit is $\frac{V_{ref}}{4} + \frac{V_{ref}}{8} + \frac{V_{ref}}{32}$, i.e. 4.0625 V. Since V_d is less than this voltage, the fifth bit is set to 0. In this way we go on comparing V_d with the sixth, seventh, eighth, ninth and tenth (LSB) bits and get the values of the bits as 1, 1, 0, 1. As we reach the tenth bit, we complete the comparison and the ADC output is finalised, and this is 0110011001 and result is 3.994140625 V for the 4V analog signal. The table gives a clearer idea of how the voltage was calculated by the ADC.

9.3 BASIC ELECTRONIC CIRCUITS

The different forms of basic electronic setup required in a nuclear experiment are shown in Figure 9.1. There are basically two processes, i.e. (a) analog and (b) digital. In the analog process the signal from the detector is processed through analog electronics until an analog-to-digital converter is used for final processing and analysis of the signal in the data acquisition system (DAQ). In the digital process, all analog electronic modules are replaced by a device known as a digitiser. In the analog process the energy and timing information of the detector are processed

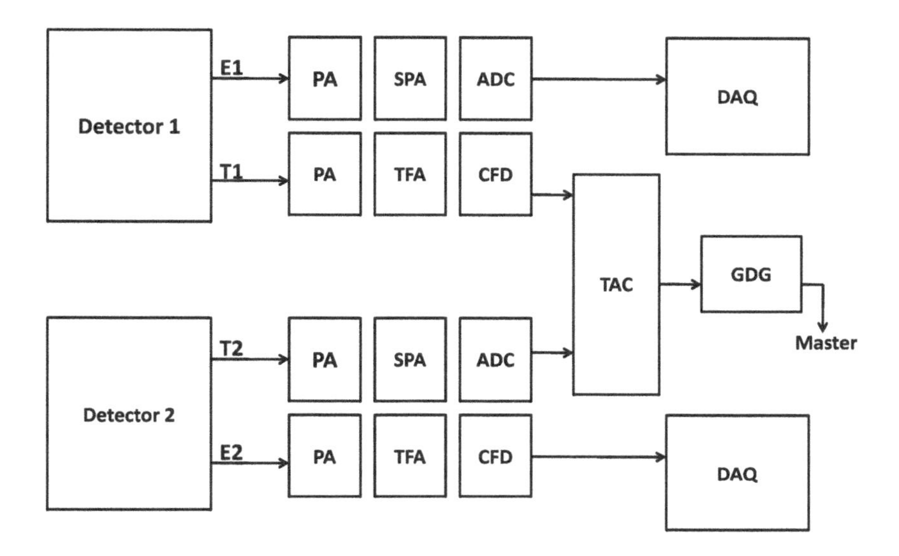

FIGURE 9.1 Basic forms of different electronic setup.

separately, whereas in the digital method time-stamped data from the detector are generated by the digitiser. The digitiser method reduces the complexity of using many modules, particularly when large number of detectors are set up in an experiment. In an analog method this is done by using high-density modules (in a single modular space there are many [usually in multiples of 16] similar modules).

In a pulse height analysis (PHA), the energy signal from the preamplifier is fed to a spectroscopy amplifier and its output to an analog-to-digital converter. The digitised pulses can be analysed in a multi-channel analyser (MCA) where each channel represents a different energy from the nuclear reaction. In a more sophisticated data acquisition software (DAQ), one can handle several such inputs from several detectors. In such a situation one may need these multiple detectors to work under some logic signal. For example, the DAQ is required to give output only when the detectors are triggered coincidentally. This is a crucial logic, because if radiations from a single event spread out in space in different directions, several detectors can detect this event placed in different orientations. If the detectors are detecting particle/radiations from a single event, then the time difference of detecting these particles in all the detectors concerned must be ideally zero or very small (usually of the order of few hundredths to tenths of a nanosecond). Therefore, a logic gate is generated from the timing signal from the preamplifier. In an analog circuit this is done as shown in Figure 9.2 for a two-detector coincidence setup. Such coincidences are also useful in particle identification setup where two or three detectors are used for particle identification.

A constant fraction discriminator (CFD) is a device that generates a logic pulse when the signal from a timing filter amplifier (TFA) crosses a discriminator level (Figure 9.3). The time difference is converted into a voltage pulse in a time-to-amplitude converter (TAC). The TAC output can be used to generate a gate signal (logic) that acts as a master gate in the DAQ. This means that DAQ will accept the ADC signal or register the energy pulse only if the master is ON. The master is ON only when the time difference between the two inputs is within the selected width (typically 100 to 200 ns). This ensures a coincidence measurement between the two detectors. If the width of TAC is very large, the rate of the master increases but the coincidence condition is compromised and many random coincidences (i.e. particles which are not from the same event but appear within the large time width) are registered. If, on the other hand, the time width selected is very small, the number of true coincidences recorded is also very small, making it difficult to understand whether sufficient coincidence events are acquired.

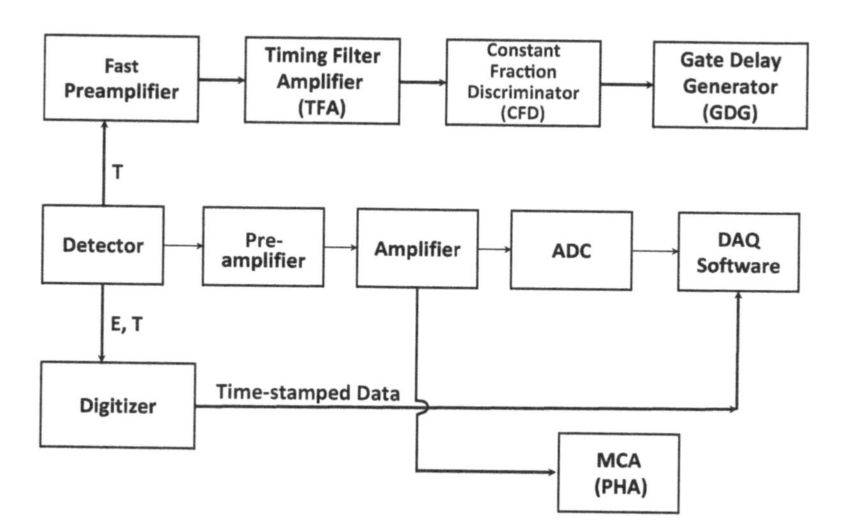

FIGURE 9.2 Electronic setup for two-detector coincidence.

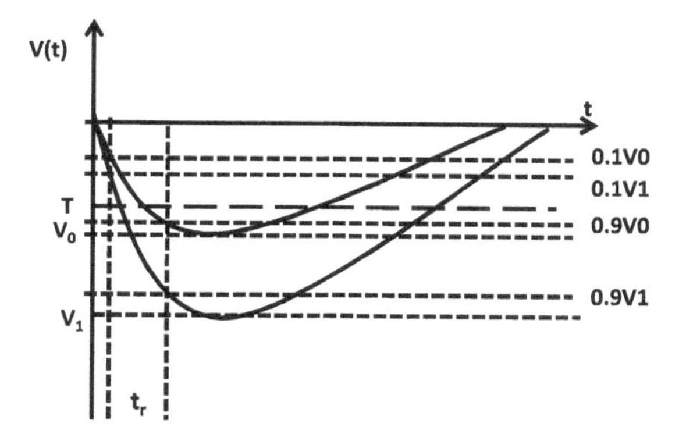

FIGURE 9.3 Leading edge discriminator.

In a practical situation there may be problems using a timing circuit. If a logic signal is generated from crossing of the leading edge of a time signal from a preamplifier, then the two pulses have the same rise time (time difference between 10% and 90% of the amplitude V0 or V1 of the two pulses). As indicated in Figure 9.3, the rise time is the time interval between points A and B, and it is the same for both pulses.

The pulses have different amplitudes, and despite having the same rise time, the two pulses cross over the discriminator threshold (bold line) at different times, generating two logic pulses at different times. Such a discrepancy is known as an amplitude walk. In order to take care of the amplitude dependence, a constant fraction discriminator (CFD) is used. This first attenuates the pulse by a fraction and adds to the input after delaying the latter by an amount necessary for the CFD signal to cross over the discriminator threshold at zero value. In a nutshell, the timing circuits play a crucial role in the electronic setup of nuclear reaction experiments.

9.4 SPECIALISED ELECTRONIC CIRCUITS

Specialised electronic setups are required in experiments that aim to achieve specific goals. One such experiment is to analyze the signals from position-sensitive detectors. Nuclear reaction cross-sections are angle dependent, and one way to measure cross-sections at different angles is to rotate the detector with respect to the beam direction or use multiple detectors at various angles. At present, position-sensitive detectors are available that provide data in a broad angular span along with larger geometric efficiency compared to using multiple detectors at various angles. Examples of position-sensitive detectors are strip solid state (silicon) detectors and multi-wire proportional (gas) counter. The process of extracting the position-sensitive signal using a 16-channel strip detector can be done either by using a charge division method or by centroid method using a multi-channel (16 channels in this case) high-density preamplifier. In the charge division method, adjacent strips are connected by equal resistances, and the induced charge on the resistive path is divided depending upon the point of injection and the total resistance travelled by the charge. If the total charge of injection is Q at a point which has a resistance of R1 on the left and R2 on the right, then after charge division, the charges that flow to the left and right are, respectively $\frac{R1.Q}{R1+R2}$ and $\frac{R2.Q}{R1+R2}$. This division of charge is carried out by a special position-sensitive module.

Alternately, one can perform this division by writing and executing a program. In the centroid method, an output from each strip is obtained in a 16-channel preamplifier. The centroid of the charge collected in the different strips determines the position of the incident radiation. Another specialised but useful circuit is the pulse shape discrimination method. This circuit is used when neutrons are detected. A neutron detector will detect both neutrons and gamma rays, which need to be separated by some electronic circuit. This is achieved by the pulse shape discrimination

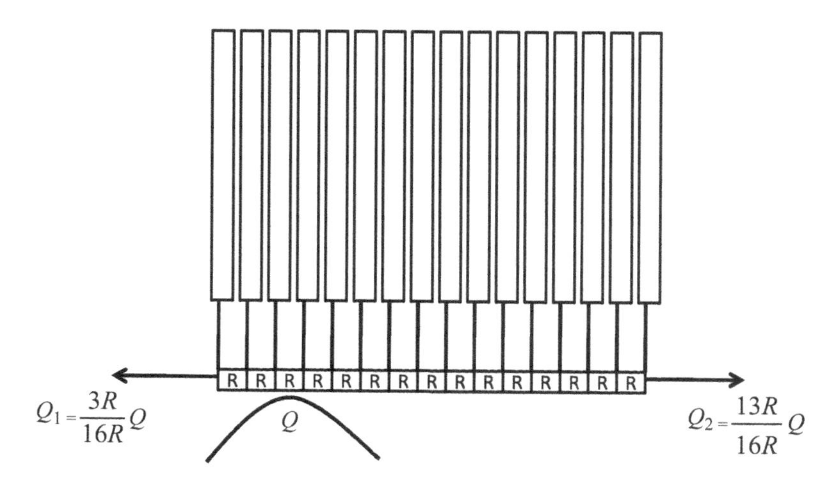

FIGURE 9.4 Position sensing by charge division method.

circuit. The neutron and gamma ray signals have different rise times, and so can be separated by their time of crossing a discriminator threshold. Alternatively, the difference in time between start and zero cross-over using a CFD can also be used for the pulse shape discrimination. Such technique is also useful for charge particle identification using CsI detectors that produce different rise times for different charged particles.

CHAPTER 10

Introduction to the Concept of Errors

Errors are an important part of experimental physics and are often overlooked by a beginner in the field. Errors can come into a measurement or appear in a calculation. It is mandatory to exhibit and discuss the sources of errors in an experiment, as well as the efforts to minimize them. A quantitative estimate of the error or uncertainty should also be quoted, which is normally expressed as a percentage. There can be various sources of errors in a completed experiment. Some are due to limitations of the instruments used and beyond the user's ability to fix them. In the following sections errors that are commonly encountered in the nuclear reaction experiments are discussed.

10.1 STATISTICAL, SYSTEMATIC AND PROGRESSIVE ERRORS

10.1.1 Statistical Errors

In any experiment, when random events are counted and recorded to determine some physical quantity that depends on this random number, an error develops in the evaluated physical quantity. The magnitude of this error is given by the square root of the number of events recorded in each measurement (from the Gaussian model). As such, this error is known as statistical error. Nuclear reaction experiments involve random events, and statistical errors need to be considered. In a nuclear measurement, the

DOI: 10.1201/9781003083863-10

detection of an event (a nuclear interaction) parameter can be guided by statistics. For example, in a measurement a scientist is recording the number of particles having energy E. As discussed in the preceding chapter on radiation detectors, this measurement is performed by radiation detectors and based upon the interaction of the radiation detected with the detector medium. The energy signal depends upon the number of charge carriers produced due to ionization. This number obviously is not exact for the same energy particle and can vary statistically, though other factors can also contribute to the variation, such as the detector's intrinsic resolution, noise and straggling. The certainty of the number of particles with energy E can be improved by counting a larger number of such particles, as the statistical error is inversely proportional to $\frac{1}{\sqrt{N}}$ where N is the number of events recorded.

From the expression for cross-section (σ) (Equation 2.5), the error in cross-section is proportional to the yield (Y) or the number of recorded random events, i.e.

$$\sigma \propto Y = k.Y \tag{10.1}$$

The error ($\Delta\sigma$) in σ from statistical model is proportional to $\sqrt{Y}$. So the absolute error is obtained by

$$\frac{\Delta\sigma}{\sigma} = \frac{\sqrt{Y}}{Y} = \frac{1}{\sqrt{Y}} \tag{10.2}$$

$$\Delta\sigma = \frac{\sigma}{\sqrt{Y}} \tag{10.3}$$

10.1.2 Systematic Errors

Systematic errors develop from the limitations in the devices used in the exprimenent. For example, if one measures length in an experiment, the device for this can be a slide rule or a slide caliper. The slide rule has an accuracy of 0.1 cm whereas a slide caliper usually has an accuracy of 0.01 cm. So using a slide caliper one can specify a length correctly to two decimal places whereas with a slide rule allows a measurement of up to one decimal place. Thus using a slide rule will result in a higher systematic error.

If the exact length of a rod is 10.75 cm, the slide rule can measure the length at best at 10.7 cm, thereby making a systematic error of .05 cm, or 0.46%. In a nuclear reaction experiment, the measurement of cross-section can have systematic errors from the uncertainties in the measured values of target thickness, solid angle and beam current. Target thickness is measured either by weighing piezoelectric crystals in thin-film evaporation devices or by alpha energy loss method. A solid angle is measured using slide calipers, and the beam current is measured using a Faraday cup and a charge integrator. All these methods have some limitations in the determination of the measured quantities and contribute to the systematic error.

10.1.3 Progressive Errors

Progressive errors occur for a quantity that depends on several other measurable parameters through some mathematical relation. For example, if the error of a quantity C which is a product of two measureable quantities A and B has to be determined, then this error propagates from the errors of A and B. Usually these errors are independent of each other, and so the error of C is a sum of quadratures of the errors of A and B, i.e. $\Delta C = \sqrt{\Delta A^2 + \Delta B^2}$.

The total error in the cross-section (σ_{total}) is therefore obtained by a sum of the squares of these errors, i.e.

$$\Delta\sigma_{total} = \sqrt{(\Delta\sigma)^2 + (\Delta x)^2 + (\Delta\Omega)^2} \qquad (10.4)$$

where the first term is the statistical error, the second term is the error in the measured thickness of the target, and the third term is the error in the measured solid angle subtended by the detector at the target centre. The target thickness if determined by weighing from the relation

$$x = \frac{m}{a} \qquad (10.5)$$

where m is the mass of the target and a is the effective area of the target. The progressive error Δx is given by

$$\Delta x = \sqrt{(\Delta m)^2 + (\Delta a)^2} \qquad (10.6)$$

The error (Δm) arises from the accuracy of the weighing balance, and the error in area comes from the error in measurement of length and breadth of the target. The error in the solid angle $\Delta\Omega$ is given by

$$\Delta\Omega = \sqrt{(\Delta s)^2 + 2(\Delta d)^2} \tag{10.7}$$

where Δs is the error in the determination of the cross-sectional area of the detector and Δd is the error in the measurement of the distance d of the detector from the target centre. As solid angle is defined as

$$\Omega = \frac{s}{d^2} \tag{10.8}$$

and the solid angle depends inversely on the square of the target-detector distance, its contribution to the total error is twice than that from the area.

10.2 CONFIDENCE LIMIT

A confidence interval is decided by how much the measured value deviates from the most probable value. Usually the confidence intervals are 1 (68%), 2 (97%) and 3σ (97%) of the counting distribution, where σ is the standard deviation and is given by the square root of the mean value in a Gaussian distribution. An example will clarify the concept. An energy measurement as discussed in detectors is always uncertain by the statistical fluctuation in the number of electron-hole (electron-ion) pairs produced from the same energy particle by interaction with the detector material. After making a large number of measurements, one can find the uncertainty usually expressed by $\sigma = \sqrt{n}$ where n is the number of counts under the peak. This is true particularly when a large number of particles (particles, holes, ions etc.) have been detected.

10.3 REDUCTION OF ERRORS IN NUCLEAR REACTION MEASUREMENTS WITH EXAMPLES

In a nuclear reaction measurement, the cross-section is the most measurable quantity. The error in the cross-section is the progressive error in terms of the errors of the measurable quantities that are present in the expression of an experimental cross-section. The yield is affected by the statistical uncertainty, whereas the thickness of the target and the solid angle present systematic uncertainties. In a measurement, if the recorded yield is 100, then the percentage error is $\frac{\Delta\sigma}{\sigma}\times 100$, i.e. $\frac{100}{\sqrt{100}}=10$. So the percentage error in this case is 10%. If the recorded yield for the same experiment is extended to 10,000, then the percentage error reduces to 1%. The errors are depicted graphically by error bars as shown in Figure 10.1. In order to increase the yield (Equation 2.5), either all or any of detection time, beam current, target thickness and/or solid angle should be increased. In a particular reaction, the cross-section determines the rate of the emissions (which cannot be changed), but the above-mentioned parameters can be altered by the experimentalist.

The measurement of energy of radiation is another important and popular detection in a nuclear reaction experiment. Except for statistical

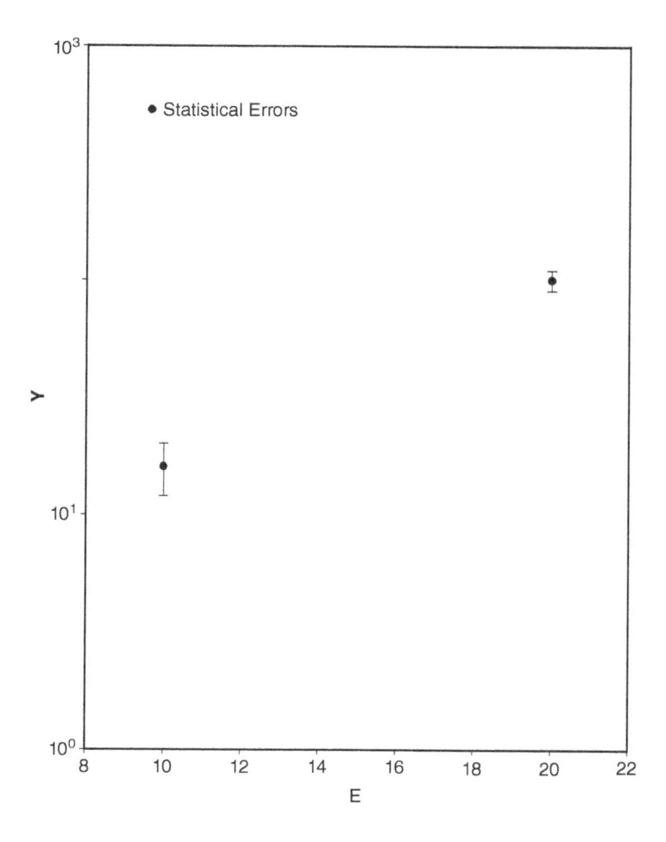

FIGURE 10.1 Reduction of statistical error.

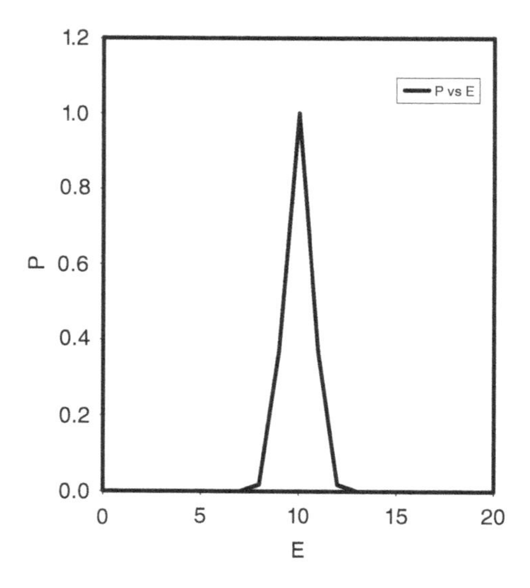

FIGURE 10.2 Uncertainty or error in energy measurement.

decays, the emission of nuclear radiation is due to transition from one discrete level to another. In such a situation the particle has a definite energy which shows up as a Gaussian peak in a pulse height distribution spectrum as shown in Figure 10.2.

As can be seen from Figure 10.2, the monochromatic nature is not reflected in the measured spectra. Instead it shows an energy spread and contributes to the errors of the measured energy. The spread or error in energy (ΔE) is measured in terms of the FWHM (full width at half maximum), and the ratio of ΔE to E times 100 is known as the resolution of the detector. The error in energy results from statistical fluctuation of the number of electron-hole (or electron-ion in case of gas detectors) pairs created by the same energy particle in the detector medium and also from the electrical noise from the detector and processing electronics.

CHAPTER 11

Theoretical Models

Theoretical models are essential to understand the underlying physics in a measurement performed in a nuclear reaction experiment. In a nuclear reaction the interactions are not exactly known and often many appropriate approximations are made that make a model of the actual situation. This is unlike, for example, Rutherford scattering used for material analysis, whose interaction and its cross-section are exactly known. The detailed theoretical derivations are not required for an experimentalist. Instead, the underlying concepts, working formulae and the importance of various parameters in these models are more important. Various computer programs in the framework of these models are available that are used to extract the physics conclusion of a measured data.

11.1 INTRODUCTION TO DIFFERENT REACTION MODELS

Nuclear reactions at low energies mentioned before can be classified mainly into direct and compound nuclear reactions. In the direct reaction part, scattering and reactions are considered. In order to calculate the cross-sections, we utilise a quantum mechanical picture. As nuclear interaction is a many-body interaction, one can, in principle, use many-body theories such as the Hartree-Fock method. However the phenomenological models are simple, popular and sufficient to analyse low-energy nuclear

DOI: 10.1201/9781003083863-11

reaction experimental data. The interaction potential used in almost all these models is a one-body optical potential that has a real and an imaginary part (similar to a complex refractive index in optics). The optical model, the statistical model, DWBA theory and coupled channel theory are some relevant models for these reactions. There is an intermediate type, known as pre-compound or pre-equilibrium reactions. The astrophysical reactions, though mainly compound nuclear in nature (with some exceptions), involve resonances. The R-matrix theory is very much suitable for studying resonance reactions.

Elastic scattering cross-sections can be calculated in the framework of the scattering theory using the optical potential. Traditionally this model is popularly known as the optical model, though in other reaction models the optical potential is also used. The optical model can only calculate elastic scattering angular distributions and total reaction cross-section. The model cannot take into consideration any change of state of the initial pair of reacting nuclei in the scattering process, i.e. it cannot calculate the inelastic cross-sections for specific states. As the total and elastic cross-sections can be calculated in the optical model, their difference yields the total reaction cross-section but not for a specific state. The distorted wave born approximation (DWBA) is suitable to calculate the inelastic and reaction cross-section. The coupled channel theory can consider the effect of one reaction process on the other, as all these processes are interlinked through the total cross-section. Thus, a coupled channel theory is a more general theory for the direct reaction community.

The statistical model is most suitable for a compound nuclear reaction that involves a continuum of levels, where quantum mechanical information about individual level is not available. However the Hauser Feshbach theory uses the quantum mechanical picture if a compound nucleus populates a discrete low-lying level of a nucleus after emission.

11.2 BASIC THEORY OF TRANSMISSION AND SCATTERING THROUGH A BARRIER

Most nuclear reaction models use a potential (interaction) concept to calculate a scattering or reaction cross-section. The potential most simply has an attractive nuclear (well) and a repulsive Coulomb (barrier) part. The incident projectile is represented by a wave function that after interaction with the potential gets either scattered from the barrier or trapped in the well, giving rise to a compound nuclear reaction. The Schrodinger equation is used to calculate the scattering and absorption probability, and relevant cross-sections can be calculated from these quantities.

The Schrödinger equation in terms of a potential V is well known, i.e.

$$\sum_i \left(E - T - V(\mathbf{r},\xi)\right)\phi_i \frac{\chi_i(r)}{r} P_l(cos\theta) = 0 \qquad (11.1)$$

The kinetic energy operator in the above is $T = -\frac{\hbar^2}{2\mu}\nabla^2 = -\frac{\hbar^2}{2\mu}\left(\frac{1}{r^2}\frac{\partial}{\partial r}\left(r^2\frac{\partial}{\partial r}\right) - \frac{L^2}{r^2}\right)$. It is to be noted that the potential is a function of the relative position vector $\mathbf{r}$, between the two interacting nuclei a and A. The wave function ϕ_i is the intrinsic wave function of the initial state consisting of a and A. Multiplying the equation with the complex conjugate of a set of final states that may be populated and integrating over the entire volume,

$$\sum_i \int d^3\xi \phi_f^* \left(E - T - V(\mathbf{r})\right)\phi_i \left(\frac{\chi_i(r)}{r}\right) P_l(cos\theta) = 0 \qquad (11.2)$$

$$\sum_i \int d^3\xi \phi_f^* \left(E + \frac{\hbar^2}{2\mu}\left(\frac{1}{r^2}\frac{\partial}{\partial r}\left(r^2\frac{\partial}{\partial r}\right) - \frac{L^2}{2\mu r^2}\right) - V(\mathbf{r})\right)\phi_i \left(\frac{\chi_i}{r}\right) P_l(\cos\theta) \qquad (11.3)$$

If the final states are in the same mass partition i.e. for elastic and inelastic scattering the intrinsic states are orthonormal i.e. $\int d^3\xi \phi_f^* \phi_i = \delta_{fi}$. Therefore, equation is simplified to

$$\left(E - \frac{\hbar^2}{2\mu}\frac{d^2}{dr^2} - \frac{\hbar^2}{2\mu}\frac{l(l+1)}{r^2}\right)\chi_f P_l - \sum_i \int d^3\xi \phi_f^* V \phi_i \chi_i(r) = 0 \qquad (11.4)$$

In V, a part can be assumed to have only radial dependence such that

$$V(\mathbf{r},\xi) = V(r) + \Delta V(\mathbf{r},\xi) \qquad (11.5)$$

and equation can be written as

$$\left(E-\frac{\hbar^2}{2\mu}\frac{d^2}{dr^2}-\frac{\hbar^2}{2\mu}\frac{l(l+1)}{r^2}-V(r)\right)\chi_f-\sum_i\int d^3\xi\phi_f^*\Delta V\phi_i\chi_i(r)=0 \quad (11.6)$$

This is the well-known coupled channel equation, as it affects the scattering solution by other channels (other than the elastic channel through the non-diagonal coupling term [integral term]).

In the simplest situation, $V(\mathbf{r},\xi) = V(r)$ and the equation reduces to the well-known radial Schrödinger equation as the last term (coupling term) vanishes,

$$\left(E-\frac{\hbar^2}{2\mu}\frac{d^2}{dr^2}-\frac{\hbar^2}{2\mu}\frac{l(l+1)}{r^2}-V(r)\right)\chi_f=0 \quad (11.7)$$

or

$$\frac{d^2}{dr^2}+\frac{2\mu}{\hbar^2}\left(E-V(r)-\frac{l(l+1)\hbar^2}{2\mu r^2}\right)\chi_f=0 \quad (11.8)$$

The equation represents a one-dimensional radial Schrödinger equation with an effective potential $V_{eff}(r)=V(r)+\frac{l(l+1)\hbar^2}{2\mu r^2}$. In case of charged nuclei undergoing nuclear reaction, $V(r)$ contains a repulsive Coulomb potential due to the positive nuclear charges in the projectile and target nuclei. The centrifugal barrier (V_{cf}) arises due to the relative angular momentum and is important at higher energies and in reactions involving heavier projectile as $l=\sqrt{2\mu k}$. The nuclear potential V_N is attractive and can be considered as a complex potential with radial dependence only so that

$$V_{eff}=-\left(U(r)+iW(r)\right)+\frac{z_a z_A e^2}{r}+\frac{l(l+1)\hbar^2}{2\mu r^2} \quad (11.9)$$

The solution of the equation yields the transmission (penetration) probability (T) through the potential barrier, i.e.

$$T=\frac{\left|\chi_f(r=R_n)\right|^2}{\left|\chi_f(r=R_c)\right|^2} \quad (11.10)$$

where R_n is the range of the nuclear potential and R_c (classical turning point) is defined by the relation

$$E = V_B\left(R_c\right) = \frac{z_a z_A e^2}{R_c} \tag{11.11}$$

and the scattering probability ($S = 1 - T$).

11.2.1 Direct Reaction Models

The solution of equation is obviously numerically done, and there are several programs available that can be used to calculate the distorted waves. The wave function χ_f is known as distorted waves that are generated due to the distortion of the plane waves by the optical potential $U + iW$. The plane waves are either the free particle ($V_{eff} = V_{cf}$) wave function in a neutron-induced reaction or a Coulomb distorted plane wave in a charged particle–induced reaction.

$$\left(\frac{d^2}{dr^2} + \frac{2\mu}{\hbar^2}\left(E - \frac{l(l+1)\hbar^2}{2\mu r^2}\right)\right)\chi_f = 0 \tag{11.12}$$

The equation transforms to a spherical Bessel differential equation if kr is substituted by ρ and E is taken common from the second term so that

$$\left(\frac{d^2}{d\rho^2} + \left(1 - \frac{l(l+1)}{\rho^2}\right)\right)\chi_f = 0 \tag{11.13}$$

The solution of the spherical Bessel equation is

$$\chi_f = A_l j_l\left(\rho\right) + B_l n_l\left(\rho\right) \tag{11.14}$$

with A_l, B_l are constants, $j_l(\rho)$ and $n_l(\rho)$ are the spherical Bessel and Neumann functions, respectively. Any second-order differential equation is solvable in terms of two independent functions. One of them is a regular solution that vanishes at the origin. In the spherical Bessel case, $j_l(\rho)$ is the regular solution as $j_l(0) = 0$. The inclusion of the Coulomb term replaces free particle equation by

$$\left(\frac{d^2}{dr^2} + \frac{2\mu}{\hbar^2}\right)\left(E - V_{cb} - \frac{l(l+1)\hbar^2}{2\mu r^2}\right)\chi_f = 0 \tag{11.15}$$

where $V_{cb} = \frac{z_1 z_2 e^2}{r}$ is the Coulomb potential expressed in terms of the number of protons z_1 and z_2 in the projectile and target, respectively, and r the distance between the centres of the interacting projectile and target. The above equation is usually rewritten by taking E common from the second term and making a substitution $\rho = kr$ ($k^2 = \frac{2\mu E}{\hbar^2}$) so that the equation now becomes

$$\left(\frac{d^2}{d\rho^2}\right) + \left(1 - \frac{2\eta}{\rho} - \frac{l(l+1)}{\rho^2}\right)\chi_f = 0 \tag{11.16}$$

This is a Coulomb differential equation. The solution of the Coulomb differential equation replaces the regular function j_l by regular Coulomb function $F_l(\eta,\rho)$ and n_l by irregular Coulomb function $G_l(\eta,\rho)$. In the case of a nuclear scattering, the complex imaginary potential prevents an analytic solution within its range contrary to when only Coulomb and no interaction is present. The solution for χ_f has two regions. The radius up to which the nuclear interaction (R_n) is present, is the inside region, and outside of that region only Coulomb term is present. The centrifugal term is present, however, as the Coulomb term in both regions. The solutions outside the nuclear range are known and are the solutions of the Coulomb equation, i.e.

$$\chi_f = A_l F_l(\eta,\rho) + B_l G_l(\eta,\rho) \tag{11.17}$$

The inside solution (left side of the equation that follows) is complicated and has to be performed numerically. Matching the logarithmic derivative of the inside and outside solutions at the nuclear range yields the unknown constants A_l and B_l, i.e.

$$\frac{\chi_f(R_n)}{\chi_f'(R_n)} = \frac{A_l F_l(\eta,\rho_n) + B_l G_l(\eta,\rho_n)}{A_l F_l'(\eta,\rho_n) + B_l G_l'(\eta,\rho_n)} \tag{11.18}$$

where $\rho_n = kR_n$. There are two unknown constants, and a judicious choice is $A_l = C_l cos\delta$ and $B_l = C_l sin\delta$. On account of the ratio taken in the above equation, a determination of δ from the above boundary condition leads to the solution. This constant δ is known as a phase shift. In case R_n is large so that asymptotic forms of the Coulomb functions are acceptable, the solutions (numerator of the right side of equation) become

$$\chi_f = C_l \cos\delta \sin\left(\rho - l\frac{\pi}{2} + \sigma_l - \eta ln(2\rho)\right) \quad (11.19)$$

$$+C_l \sin\delta \cos\left(\rho - l\frac{\pi}{2} + \sigma_l - \eta ln(2\rho)\right) \quad (11.20)$$

$$= C_l \sin\left(\rho - l\frac{\pi}{2} + (\sigma_l + \delta) - \eta ln(2\rho)\right) \quad (11.21)$$

The purpose of any nuclear reaction model is to provide an expression to calculate the cross-section. In the scattering process, where there is no change in the internal state of the two interaction nuclei and so the intrinsic structure wave functions can be ignored or does not appear, the incident plane wave or Coulomb distorted plane wave after interaction with the target a part of the incident flux can either move on uninterrupted or a part gets scattered radially as a spherical wave. The amplitude of this scattered spherical wave is known as the scattering amplitude *f* and is considered to be generally dependent on the angle at which the wave is scattered. Therefore, the total wave function after interaction is the incident wave and the scattered wave, i.e.

$$\chi_f = \chi_{in} + f(\theta)\frac{e^{i\rho - \eta ln 2\rho}}{k}\rho \quad (11.22)$$

The first term in right side of the above equation is the incident wave moving along the z direction, and the second term is a spherical wave moving out radially from the interaction point (at the target). These two waves at a large distance from the target are considered to be non-interfering, and the above relation is acceptable. The incident wave is a Coulomb wave and can be written as

$$\chi_{in} = \sum_l i^l (2l+1) F_l(\eta, \rho) P_l \quad (11.23)$$

The above expression is the same as a partial wave expansion of a plane wave except the exponential term introducing Coulomb phase factor σ.

In the asymptotic limit, the expression for an incident wave can be written by replacing the regular Coulomb function $F_l(\eta,\rho)$ with its asymptotic form, i.e.

$$\chi_{in} = \sum_l i^l (2l+1)\sin\left(\rho - l\frac{\pi}{2} + \sigma_l - \eta ln2\rho\right)P_l \tag{11.24}$$

or

$$\chi_{in} = \frac{1}{2i\rho}\sum_l i^l (2l+1)\left(e^{i\Theta} - e^{-i\Theta}\right)P_l \tag{11.25}$$

where $\Theta = \rho - \frac{l\pi}{2} - \sigma_l(\eta) - \eta ln(2\rho)$. Similarly, the total wave function is also corrected by the same phase factor and can be written as

$$\chi_f = \sum_l C_l \frac{e^{i(\delta+\Theta)} - e^{-i(\delta+\Theta)}}{2i\rho}P_l \tag{11.26}$$

Inserting the expression of total wave after scattering (χ_f) on the left side of Equation 11.23 and equating the coefficients of $e^{-i(\rho - \eta ln2\rho)}$

$$C_l = i^l (2l+1)e^{i\delta} \tag{11.27}$$

and then the coefficients of $e^{i(\rho - \eta ln2\rho)}$ on the right side of Equation 11.23 is

$$\frac{1}{2i\rho}\sum_l i^l (2l+1)e^{i\left(\sigma - l\frac{\pi}{2}\right)}P_l + f(\theta)\frac{k}{\rho} \tag{11.28}$$

and from the left side of Equation 11.23 is

$$\frac{1}{2i\rho}\sum_l i^l (2l+1)e^{i\delta}e^{i\left(\sigma - l\frac{\pi}{2} + \delta\right)} \tag{11.29}$$

Equating 11.28 and 11.29 gives the nuclear plus Coulomb scattering amplitude as

$$f(\theta) = \frac{1}{2ik}\sum_{l}(2l+1)e^{i\sigma}\left(e^{2i\delta} - 1\right)P_l \tag{11.30}$$

The elastic scattering cross-section is obtained in terms of the scattering amplitude as

$$\frac{d\sigma}{d'} = \left|f(\theta)\right|^2 \tag{11.31}$$

The reaction cross-section is obtained by the absorbed flux in a nuclear interaction, whereas in a scattering process there is no reduction of flux. The scattering theory ignores the structure of the nuclei and therefore cannot calculate the cross-section of a particular reaction such as inelastic scattering resulting in a projectile or target excitation to a particular state or a transfer reaction. Such reactions are addressed by the direct reaction theory such as the coupled channel formalism (see the previous section). However, the scattering theory can predict the total reaction cross-section as

$$\sigma_R = \frac{|R_{out}(r)| - |R_{inc}(r)|}{F_a} \tag{11.32}$$

In the above definition, the difference between the incoming and incident flux must be noted. The incoming and outgoing flux correspond to the spherical waves after the interaction where the amplitude of the outgoing wave is changed due to the interaction by the phase shift or S-matrix (S_l). The incident wave is a plane wave that can be made up of incoming and outgoing spherical waves of equal amplitude. The radial incoming (F_{inc}) and outgoing (F_{out}) flux are obtained from the following (Equation 11.26), i.e.

$$\chi_f = S_l u_{out} - u_{inc} \tag{11.33}$$

where $S_l = e^{2i\delta}$ and u_{out} and u_{inc} are given by

$$u_{inc} = -\sum_{l}\frac{(2l+1)}{2i\rho}e^{-i\rho'}P_l \tag{11.34}$$

and

$$u_{out} = \sum_{l} \frac{(2l+1)}{2i\rho} S_l e^{i\rho'} P_l \tag{11.35}$$

where $\rho' = \rho + \sigma - \eta ln2\rho$. The fluxes are obtained from

$$F_{inc}(r,\theta) = \frac{\hbar}{2mi}\left(u_{inc}\frac{\partial u_{inc}^*}{\partial r} - u_{inc}^*\frac{\partial u_{inc}}{\partial r}\right) \tag{11.36}$$

$$R_{inc}(r) = -\frac{\hbar}{2mi}\sum_{l,l'}\frac{(2l+1)(2l'+1)}{4i\rho^2}e^{i(\rho_{l'}' - \rho_l')}$$

$$\left(1 - \frac{\eta}{\rho}\right) \times \int P_l P_{l'}^* \rho^2 d\Omega \tag{11.37}$$

$$F_{out}(r,\theta) = \frac{\hbar}{2mi}\left(u_{out}\frac{\partial u_{out}^*}{\partial r} - u_{out}^*\frac{\partial u_{out}}{\partial r}\right)$$

$$R_{out}(r) = \frac{\hbar}{2mi}\sum_{l,l'}\frac{(2l+1)(2l'+1)}{4i\rho^2}e^{i(\rho_{l'}' - \rho_l')}$$

$$\left(1 - \frac{\eta}{\rho}\right) \times S_l S_{l'}^* \int P_l P_{l'}^* \rho^2 d\Omega \tag{11.38}$$

The orthogonality relation for Legendre polynomials, i.e.

$$\int P_l P_{l'}^* dcos(\theta) = \frac{2}{2l'+1}\delta_{ll'} \tag{11.39}$$

reduces the rates of the outgoing and incoming flux, and with the incident flux being $\frac{\hbar k}{m}$, the reaction cross-section is given by

$$\sigma_R = \frac{\sum_l \frac{\hbar(2l+1)}{4mk}\left(1 - |S_l|^2\right)4\pi}{\frac{\hbar q}{m}} \tag{11.40}$$

$$= \frac{\pi}{k^2} \sum_l (2l+1)\left(1-|S_l|^2\right) \tag{11.41}$$

with the term $\left(1-\frac{\eta}{\rho}\right)$ leading to 1 at large ρ. The calculation of reaction cross-section of specific channel is best carried out by the coupled channel formalism. Three fundamental direct reactions needs to be understood: namely, inelastic scattering, transfer and breakup reactions can be calculated from this theory. In fact, the effect of one type of these reactions can also be studied using this theory. The inelastic scattering and transfer reactions both populate discrete quantum states, and the application of relevant potential in the coupled channels equation leads to the calculation of the cross-section. The breakup reaction is a little different in the sense that it produces three nuclei in the final state and connects to the energy continuum rather than to a definite energy level. In order to apply quantum mechanical principles, therefore, the breakup continuum is discretised.

Inelastic scattering changes the state of any or both of the reacting nuclei a and A. As an example, an inelastic projectile excitation of a nucleus ^{17}O due to interaction with ^{208}Pb can be considered. As ^{17}O can be thought of a p + ^{16}O system with proton as a valence particle and ^{16}O as core, this nucleus is very likely to undergo a single-particle excitation. This means the scattering process changes the proton orbital, thereby changing the state of ^{17}O from ground state to some excited state. This is expressed as

$$^{17}O_{gs} + {}^{208}Pb \rightarrow {}^{17}O_{*} + {}^{208}Pb \tag{11.42}$$

The change in state of ^{17}O from its ground state to the excited state is brought about by three interactions. These are best understood by considering a p+^{16}O picture of ^{17}O where p is the valence particle that is disturbed by the inelastic scattering and ^{16}O is the core. In doing this, the many-body interactions in the nucleus ^{17}O are simplified by assuming a two-body picture with a single-particle potential (similar to reducing the two-body problem of a helium atom to a one-body system). The interactions involved are the real binding potential V_{cv} for the p+^{16}O system in its ground or excited state and the optical potentials for the ^{16}O + ^{208}Pb system. The total wave function therefore has only two states, i.e. the elastic state ($\phi_1\chi_1$) and the state with the projectile excited ($\phi_2\chi_2$). Two distorted wave states lead to two coupled equations from the general expression, i.e

$$\left(E - \frac{\hbar^2}{2\mu}\frac{d^2}{dr^2} - \frac{\hbar^2}{2\mu}\frac{l(l+1)}{r^2} - V(r)\right)\chi_1 = \Delta V_{12}\chi_2(r) \tag{11.43}$$

$$\left(E - \frac{\hbar^2}{2\mu}\frac{d^2}{dr^2} - \frac{\hbar^2}{2\mu}\frac{l(l+1)}{r^2} - V(r)\right)\chi_2 = \Delta V_{21}\chi_1(r) \tag{11.44}$$

where

$$\Delta V_{ij} = \int \phi_i \Delta V \phi_j d\xi \tag{11.45}$$

The non-diagonal matrix element of the interaction matrix is obtained by using bound state wave functions. In case of more states, the number of coupled equations increases and calculations have to be done in an iterative way. In the above simple case, one can consider the coupling potential V_{12} very weak, i.e. $V_{12} \approx 0$. This enables the solution of χ_1 from the calculated values of ϕ_1 and ϕ_2 that are obtained by solving the Schrödinger equation using V_{cv} as the interaction potential.

$$\left(E - \frac{\hbar^2}{2\mu}\frac{d^2}{dr^2} - \frac{\hbar^2}{2\mu}\frac{l(l+1)}{r^2} - V(r)\right)\chi_1 \approx 0 \tag{11.46}$$

Putting this solution of χ_1 in the second equation yields the value of χ_2. These solutions, if used to evaluate the T-matrix and the cross-section, are the first-order distorted wave born approximation (DWBA) solutions. If the first-order χ_2 is inserted in the first equation, an improved value of χ_1 is obtained which, when put in the second equation, gives further improved χ_2. The latter are known as second-order DWBA solutions. In this way a large number of DWBA solutions can be tried, but for most practical purposes first-order DWBA is good enough.

Transfer reaction cross-sections can be also calculated using the coupled channel equations. In this case, again, a practical example is considered for better understanding, namely the alpha transfer reaction

$$^{6}\text{Li} + {}^{12}\text{C} \rightarrow d + {}^{16}\text{O} \tag{11.47}$$

In the alpha transfer process, the projectile ^{6}Li is considered to be made up of α + d (a very good assumption, as alpha has a very low separation energy in ^{6}Li) and after interaction with ^{12}C, the alpha particle is transferred to it, forming ^{16}O (=α + ^{12}C) in different states. In this process, the transferred particle (i.e. alpha) is known as the valence (v) particle, and d and ^{12}C act as the initial (c_1) and final c_2 core nuclei. The Hamiltonian for this three-body system can be written in terms of the Jacobi coordinates (r, r', R, R') as

$$h = t(r) + t(R) + v(r) + V(Rc_1c_2) + v(r') \tag{11.48}$$

or equivalently,

$$h = t(r') + t(R') + v(r) + V(Rc_1c_2) + v(r) \tag{11.49}$$

The consideration of projectile bound state (i.e. before the interaction) in the first Hamiltonian gives the Hamiltonian in its “prior” form and the second one in “post” form, i.e. after the interaction. The prior- and post-Hamiltonians can be rewritten with a single-particle approximation as

$$h = t(r) + v(r) + t(R) + V(R) + \Delta V \tag{11.50}$$

and

$$h = t(r') + v(r') + t(R') + V(R') + \Delta V' \tag{11.51}$$

where $V(R)$ and $V(R')$ are optical potentials for the ^{6}Li + ^{12}C and d + ^{16}O channels, respectively. Therefore, the residual interaction in each case is

$$\Delta V = v(r') + V(Rc_1c_2) - V(R) \tag{11.52}$$

$$\Delta V' = v(r) + V(Rc_1c_2) - V(R') \tag{11.53}$$

The last two terms $V(Rc_1c_2)$ (core-core potential) and $V(R)$ or $V(R')$, the entrance or exit channel optical potentials in each expression, are complex potentials and are expected to be very close to each other. The above interaction can again be used in the coupled channel equations as in the case of inelastic scattering, and transfer cross-sections can be calculated. The only difference is the radial coordinates in the set of two coupled channel equations in the case we consider as in inelastic scattering. The first two terms in the Hamiltonian act on the intrinsic states, and the next two terms generate the distorted waves in the initial and final channel. The coupled channel equations for transfer are written as

$$\left(E-\varepsilon_i-\frac{\hbar^2}{2\mu}\frac{d^2}{dr^2}-\frac{\hbar^2}{2\mu}\frac{l(l+1)}{r^2}-V(r)\right)\chi_1=\Delta V\chi_2(r) \quad (11.54)$$

$$\left(E-\varepsilon_f-\frac{\hbar^2}{2\mu}\frac{d^2}{dr'^2}-\frac{\hbar^2}{2\mu}\frac{l(l+1)}{r'^2}-V(r')\right)\chi_2=\Delta V'\chi_1(r') \quad (11.55)$$

The solutions can again be first- or higher-order DWBA or full coupled channel calculations. The breakup reaction involves a transition of either the target or the projectile from an initial bound state (discrete state) to an unbound state (described by a continuum of levels). The most common breakup reaction encountered in nuclear physics is projectile breakup of a nucleus in which constituents have a low separation energy. The cross-section for breakup can also be obtained from coupled channel theory as described for inelastic and transfer reactions. However, the wave function of the projectile is a combination of discreet bound and continuum unbound states. Contrary to the simplest case of the ground state and a bound excited state for projectile inelastic excitation, a large number of levels are present as the continuum besides the ground state in the breakup case. So the wave function is

$$\phi_i(r)\chi_i(R)+\int d^3k\phi_i(r,k)\chi_i(R,K) \quad (11.56)$$

The second term in the above equation represents the continuum and cannot be considered in a quantum mechanical framework where one works with discrete states. There are several discretisation methods of the continuum states, and this leads to the continuum discretised coupled channel (CDCC) approximation.

11.3 COMPOUND NUCLEAR REACTION MODEL

$$\sigma(\varepsilon,\theta)=\sum_{0}^{l_c}\sum_{|s_a-s_A|}^{(s_a+s_A)}\sum_{|J=l-s|}^{(l+s)}\sigma_{ls}^{J}(E)P^{J}(\varepsilon,\theta) \quad (11.57)$$

where the $\sigma_{ls}^{J}(E)$ is the fusion cross-section at a particular angular momentum populated for the compound nucleus, i.e.

$$\sigma_{ls}^{J}(E)=\pi\lambda^{2}\sum_{m_s,M}(2l+1)T_{ls}^{J}(E)\left|C_{0m_sM}^{lsJ}\right|^{2}\delta_{M,m_s} \quad (11.58)$$

$$=\pi\lambda^{2}\sum_{m_s}(2l+1)T_{ls}^{J}(E)\left|C_{0m_sm_s}^{lsJ}\right|^{2} \quad (11.59)$$

The decay probability of a particle of type x from the compound nucleus is $P^J(\varepsilon,\theta)$ and is given by

$$P^{J}(\varepsilon,\theta)=\sum_{s'=|s_b-s_B|}^{(s_b+s_B)}\sum_{l'=|J-s'|}^{J+s'}P_{l's'}^{J}(\varepsilon,\theta) \quad (11.60)$$

$$=\sum_{l',s'}\frac{T_{xl's'}^{J}(\varepsilon)}{\sum_{x,\varepsilon}T_{xl's'}^{J}(\varepsilon)}\sum_{m_{l'}m_{s'}}\left|C_{m_{l'}m_{s'}m_s}^{l's'J}\right|^{2}\left|Y_{l'm_{l'}}(\theta,\phi)\right|^{2}\delta_{m_{l'},\Delta m_s} \quad (11.61)$$

$$=\sum_{l',s'}\frac{T_{xl's'}^{J}(\varepsilon)}{\sum_{x,\varepsilon}T_{xl's'}^{J}(\varepsilon)}\sum_{m_{s'}}\left|C_{\Delta m_sm_{s'}m_s}^{l's'J}\right|^{2}\left|Y_{l'\Delta m_s}(\theta,\phi)\right|^{2} \quad (11.62)$$

$$=\sum_{l',s'}\frac{T_{xl's'}^{J}(\varepsilon)}{\sum_{x,\varepsilon}T_{xl's'}^{J}(\varepsilon)}\sum_{m_{s'}}\left|C_{\Delta m_sm_{s'}m_s}^{l's'J}\right|^{2}Y_{l'\Delta m_s}(\theta,\phi)Y_{l'\Delta m_s}^{*}(\theta,\phi) \quad (11.63)$$

where $\Delta m_s = m_s - m_{s'}$. The product of a spherical harmonic and its complex conjugate can be decomposed in the following way:

$$Y_{l'\Delta m_s}(\theta,\phi)Y_{l'\Delta m_s}^{*}(\theta,\phi)=Y_{l'\Delta m_s}(\theta,\phi)(-1)^{\Delta m_s}Y_{l'-\Delta m_s}(\theta,\phi) \quad (11.64)$$

$$=(-1)^{\Delta m_s}Y_{l'\Delta m_s}(\theta,\phi)Y_{l'-\Delta m_s}(\theta,\phi) \quad (11.65)$$

Using the expansion for product of spherical harmonics,

$$Y_{l_1m_1}(\theta,\phi)Y_{l_2m_2}(\theta,\phi)=\sum_{LM}\sqrt{\frac{(2l_1+1)(2l_2+1)}{4\pi(2L+1)}}Y_{LM}(\theta,\phi)C_{m_1m_2M}^{l_1l_2L}C_{000}^{l_1l_2L} \quad (11.66)$$

the product of spherical harmonics can be written as

$$(-1)^{\Delta m_s} Y_{l'\Delta m_s}(\theta,\phi) Y_{l'-\Delta m_s}(\theta,\phi)$$

$$= (-1)^{\Delta m_s} \sum_{LM} \sqrt{\frac{(2l'+1)(2l'+1)}{4\pi(2L+1)}}$$

$$\times Y_{LM} C^{l'l'L}_{\Delta m_s - \Delta m_s M} C^{l'l'L}_{000} \delta_{0M} \tag{11.67}$$

$$= (-1)^{\Delta m_s} \sum_{L} \sqrt{\frac{(2l'+1)(2l'+1)}{4\pi(2L+1)}}$$

$$\times Y_{L0} C^{l'l'L}_{\Delta m_s - \Delta m_s 0} C^{l'l'L}_{000} \tag{11.68}$$

$$= (-1)^{\Delta m_s} \sum_{L} \sqrt{\frac{(2l'+1)(2l'+1)}{4\pi(2L+1)}}$$

$$\times \sqrt{\frac{(2L+1)}{4\pi}} P_L(cos\theta) C^{l'l'L}_{\Delta m_s - \Delta m_s 0} C^{l'l'L}_{000} \tag{11.69}$$

$$= (-1)^{\Delta m_s} \frac{(2l'+1)}{4\pi} \sum_{L} P_L(cos\theta)$$

$$\times C^{l'l'L}_{\Delta m_s - \Delta m_s 0} C^{l'l'L}_{000} \tag{11.70}$$

$$P^J(\varepsilon,\theta) = \sum_{l',s'} \frac{T^J_{xl's'}(\varepsilon)}{\sum_{x,\varepsilon} T^J_{xl's'}(\varepsilon)} \sum_{m_{s'}} \left| C^{l's'J}_{\Delta m_s m_{s'} m_s} \right|^2 \tag{11.71}$$

$$\times (-1)^{\Delta m_s} \frac{(2l'+1)}{4\pi} \sum_{L=0,\Delta L=even}^{2l'} P_L(cos\theta) C^{l'l'L}_{\Delta m_s - \Delta m_s 0} C^{l'l'L}_{000} \tag{11.72}$$

Since the Clebsch–Gordon Coefficient $C^{l'l'L}_{000}$ vanishes for odd L values, the compound nuclear cross-sections shows a symmetry at about a 90-degree angle of emission, as do the even values of $P_L(cos\theta)$. At very low energies only $l' = 0$ waves are possible. In such a situation the angular distribution shows an isotropic (independent of angle) nature, as then only $L = 0$ values are possible and P_0 is independent of θ. The above formalism holds good for discrete state population of the residual nucleus B. In compound nuclear emissions, residual nucleus can be populated to a continuum of energy levels. In such a situation the probability of decay at an energy ε is additionally determined by the nuclear level density of the residual nucleus $\rho(U)$ at the residual excitation energy $U = E_c - s_x - \varepsilon$, i.e.

$$P^J(\varepsilon,\theta) = \sum_{l',s'} \frac{T^J_{xl's'}(\varepsilon)\rho(U)}{\sum_{x,\varepsilon} T^J_{xl's'}(\varepsilon) + \sum_x \int_0^{E_c - s_x} T^J_{xl's'}(\varepsilon)\rho(U)dU}$$
$$\times \sum_{m_{s'}} \left| C^{l's'J}_{\Delta m_s m_{s'} m_s} \right|^2 (-1)^{\Delta m_s} \frac{(2l'+1)}{4\pi}$$
$$\times \sum_{L=0,\Delta L=even}^{2l'} P_L(cos\theta) C^{l'l'L}_{\Delta m_s -\Delta m_s 0} C^{l'l'L}_{000} \tag{11.73}$$

The transmission coefficients $T^J_{ls}(\varepsilon)$ or $T^J_{l's'}(\varepsilon)$ are determined from $a + A$ and $b + B$ optical potentials, respectively determined from fitting the elastic scattering data.

11.4 THEORIES OF ASTROPHYSICAL REACTIONS

In astrophysical reactions there are two types of measurements and consequent analysis. One is direct measurement of the low-energy astrophysical reaction. Most astrophysical reactions show some distinctive features from general nuclear reaction study: (a) relative energy of reactants much lower than the Coulomb barrier; (b) light charged particles are reactants; (c) mainly gamma rays are observed (except in some cases alpha particles are emitted); and (d) compound nuclear reactions populate resonance states.

All these features are well treated by a theory known as the R-matrix formalism. The R-matrix is defined as

$$R = \frac{1}{a} \frac{u_{in}(a)}{u'_{in}(a)} \tag{11.74}$$

where a is the nuclear radius beyond which there is no effect of the nuclear interaction and u_{in} are the radial wave function inside the nuclear periphery and $u'_{in}(a)$ is its derivative. All $u_{in}(r)$ do not generate a resonance. But clearly if $u'_{in}(a) = 0$, it indicates a divergence in the R-matrix and hence a pole in the S-matrix (that calculates the cross-section). The total wave function in the framework of the scattering theory inside the range of nuclear interaction along with Coulomb interaction (present at both small and large separation of the reactants) gives

$$R = \frac{1}{a}\left(\frac{F\cos\delta + G\sin\delta}{F'\cos\delta + G'\sin\delta}\right) \tag{11.75}$$

$$= \frac{1}{a}\left(\frac{F + G\tan\delta}{F' + G'\tan\delta}\right) \tag{11.76}$$

$$\tan\delta = \frac{1}{i}\frac{e^{i\delta} - e^{-i\delta}}{e^{i\delta} + e^{-i\delta}} \tag{11.77}$$

$$= \frac{1}{i}\frac{e^{i2\delta} - 1}{e^{i2\delta} - 1} \tag{11.78}$$

$$= \frac{1}{i}\frac{S - 1}{S + 1} \tag{11.79}$$

The S-matrix expression in terms of the R-matrix becomes

$$S = \frac{(G - iF) - aR(G' - iF')}{(G + iF) - aR(G' + iF')} \tag{11.80}$$

A resonance occurs at a S-matrix pole, i.e. when the R-matrix is given by the purely outgoing Coulomb wave, i.e. $R = \frac{1}{a}\frac{H^{+}}{H'^{+}}$.

11.5 PRE-COMPOUND REACTION MODELS

In reactions where incident energy is high (10 MeV/A or more), a phenomena known as a pre-compound nuclear reaction is observed. In this reaction process particles or gamma rays are emitted from a composite nucleus that is not fully equilibrated. Pre-compound reactions are therefore sometimes known as pre-equilibrium reactions. A very nice and simple evolution of the composite (not compound) nucleus to the final equilibrated compound system and pre-compound particle emission in the process is described by the exciton model proposed by J.J. Griffin. In the exciton model, a nucleon interacts with a target nucleus in which the nucleons are arranged in a equidistant spacing model. The incident nucleon is removed from the entrance channel if it exchanges energy with a nucleon in the target nucleus. By this interaction, if a nucleon is excited above the Fermi level, a hole is correspondingly excited below the Fermi level. Thus the projectile nucleon and the excited nucleon and hole from the target forms a two particle and one hole, i.e. three-exciton level. With each such interaction the exciton number increases by one particle and one hole (i.e. two excitons). In this way the exciton number increases as the composite nucleus gradually thermalises when there is a large number of excitons and energy per nucleon has decreased. If, however, a particle is emitted when the exciton number is low, then they are of much higher energy and are termed pre-equilibrium emissions.

11.6 INTRODUCTION TO SOME AVAILABLE PROGRAMS

There are several available free computer programs on different internet sites that are useful for nuclear reaction model calculations. These are reliable programs, and directions to use them are provided on the websites hosting them. Some of them are online versions, meaning that no physical installation on one's workstation is requires. Some of these programs are mentioned here.

The **Nuclear Reaction Video** (http://nrv.jinr.ru/nrv/) project website provides an online program to calculate two- and three-body kinematics. There are also programs to calculate cross-sections for different reactions.

PACE4 (http://lise.nscl.msu.edu) is a statistical model program that can be used mainly for heavy ion-induced compound nuclear emission cross-section calculations. Alpha, protons and neutron emission spectra and angular distributions can be calculated. The gamma emission cross-sections can be also calculated using this program.

TALYS (https://tendl.web.psi.ch/tendl_2019/talys.html) is a statistical model code but only for calculations with light ions projectiles like protons, neutrons, alpha, helions, tritons and deuterons. Any particle in the exit channel and gamma ray emission cross-sections can be calculated.

The program calculates both energy spectra and angular distributions and is suitable for very low energy reactions.

CASCADE (http://atomki.hu) is another statistical model code that calculates compound nuclear emission spectra for both light and heavy ions with all different types of emitted particles.

FRESCO (http://www.fresco.org.uk) is a versatile program that calculates elastic, inelastic and direct reaction cross-sections (transfer and breakup reaction) both in coupled channel and single-particle mode.

AZURE2 (https://azure.nd.edu) is a R-matrix program useful for making calculations for resonance reactions. The program is extremely useful for fitting experimental data and also provides extrapolation to energies where measurements are not possible. This is an indispensable program in nuclear astrophysics studies.

Index

Page numbers in italic indicate figures.

A

accelerators, 9, 41–42, 51–54, 57, 59, 61
active shielding, 50
amplifier, 94–95, 97–98
analog pulse, 95–98
analog to digital converter, 95–96, 98
angular distribution, 26, 29, 34, 43, 45, 77, 110, 125, 127
astrophysical S-factor, 39

B

beam current, 12, 38, 48–49, 60–61, 105, 107
Bethe–Bloch relation, 42
binding energy, 3–4, *4*, 33, 86
Bragg Curve detector, 43, 48
breakup reaction, 14, 26–27, 30, 34, 119, 122, 128

C

centre of mass frame, 15–16, 59
charge division method, 100, *100*
charged particle detector, 42–44, 48
charge state, 12, 46, 48, 52–53, 58, 61, 76–77, 82
cluster structure, 5
Cockroft–Walton, 51, 53, *55*, 57
collimator, 44
compound nucleus, 27–28, 32–41
Compton scattering, 77–78
confidence interval, 106
constant fraction discriminator, 98–99
Coulomb differential equation, 114
coupled channel, 110, 117, 119–122, 128
cross-section, 60–61, 77, 99, 104–107, 109–111, 115, 117–122, 124, 126–128
cyclic accelerator, 51, 57
cyclotron, 51, 57–59

D

DC accelerator, 41, 51
detector capacitance, 91–92, 94–95
diffusion pump, 66, 69, *69*
digital pulse, 96–97
direct reaction, 26–32, 34, 109–110, 113, 117, 119, 128
distorted wave Born approximation, 110, 113, 119–121
doorway states, 28

E

Einzel lens, 61
elastic scattering, 40, 45, 110–111, 117, 119, 121, 125
electromagnetic lens, 42
energy
    resolution, 79, 84
    spectrum, 26, 31, 34

errors, 103–104
exclusive measurement, 14

F

Fermi's golden rule, 13
flux, 11, 13, 115, 117–118
focal point detector, 42
fore vacuum, 65
fusion, 6, 27–28, 32, 35–39, 122

G

gamma emission, 6, 27, 36, 38, 49, 127
Gamow energy, 36–39
gas detector, 78–80, *80*, 81–83, 86
geometric efficiency, 42, 77, 100
gravitational energy, 35

H

Hauser–Feshbach theory, 110

I

impedance matching, 91, 95
inclusive measurement, 14
inelastic scattering, 26–29, 111, 117, 119, 121
intrinsic efficiency, 77–78
inverse kinematics, 21
Inverted magnetron gauge, 71
ion impanted detector, 86
ionization chamber, 81
ion source, 48, 52–53, 57, 59

J

Jacobian, 16, 18

K

knockout reaction, 27, 30–31
K of Cyclotron, 58–59

L

laboratory frame, 15–16, 19, 22
laminar flow, 65, 68
leak valves, 72
least significant bit, 96
lithium drifted detector, 86, 88
loosely bound nuclei, 27
Lorentz force, 48, 57

M

magic nucleus, 5
magnetic spectrometers, 42, 48
molecular flow, 65, 68
Mosley's experiment, 1
most significant bit, 96
multi-detector array, 42–43, 48
multi-wire chamber, 48, 83, *83*, 100

N

NaI (Tl) detector, 89
neutron, 2–9, 12, 26–27, 32–33, 35–36, 41–42, 48–49, 51, 59–60, 75, 77, 100–101, 113, 127
  detector, 77, 100
  generators, 59–60
nuclear mass, 3
nuclear reaction programs, 128
nucleosynthesis, 36

O

Ohmic junction, 87, *88*
optical model, 110
orbitting reaction, 27

P

partial cross-section, 12–14
passive shielding, 50
Pelletron, 53–54
Penning gauge, 70
phase space, 13

photoelectric effect, 77–78
Pirani gauge, 70–71
planetary model of atom, 2–3
Plum Pudding model, 1
preamplifier, 91–92, 94–95, 97
precompound reaction, 28, 34
progressive error, 105
projectile breakup, 27, 30, 122
proportional counter, 48, 81–82

Q

quadrupole, 49, 60–61
quasielastic scattering, 27
Q-value, 36

R

radioactive, 6, 9, 26–27, 49–50, 79
random coincidence, 98
random events, 103–104
reaction rate, 36–37
R-matrix theory, 110, 125–126, 128
roots pump, 66, 68
rotary pump, *66–67*, 68, 70
rotor, 67, 70
rp-process, 36
r-process, 36
Rutherford scattering, 45–46, 109
Rutherford's experiment, 1

S

scattering chamber, 15, 42–45
Schottky junction, 86, *87*
scintillation detector, 75, 84, 88
scroll pump, 66, *68–69*
semiconductor detector, 84–88, 92, 94
separation energy, 5, 30, 120, 122
separators, 42, 48
shielding of detectors, 38, 49–50
single wire detector, *82*, 84
solid angle, 12–14
s-process, 36
statistical error, 103, *107*
statistical gamma rays, 27
stator, 70
stripper canal, 53
stripper foil, 53
surface barrier detector, 45, 86–87
systematic error, 104–105

T

tandem acclerator, 52, *52*
target breakup, 27
terminal voltage, 53–54, 94
thermal energy, 35, 84
three body kinematics, 22–23, 127
time of flight, 44, 47
time resolution, 47
time to amplitude converter, 98
total cross-section, 12–14, 30, 110
transfer reaction, 14, 26–27, 29–30, 117, 119–120, 122
transverse field, *82*
turbomolecular pump, 65, 68, 70, *71*, 73
two-body kinematics, 19, 46

V

vacuum valves, 72–73
Van de Graaff, 51, 53, *54*
viscous flow, 68

W

waiting point nuclei, 36
Weiskopf–Ewing theory, xiii

Y

yield, 12, 17–18, 22, 60, 77, 104, 106–107, 110, 112, 114, 120